# TEACHING PROFESSIONAL
# AND TECHNICAL COMMUNICATION

# TEACHING PROFESSIONAL AND TECHNICAL COMMUNICATION

*A Practicum in a Book*

*Edited by*
**TRACY BRIDGEFORD**

**UTAH STATE UNIVERSITY PRESS**
*Logan*

© 2018 by University Press of Colorado

Published by Utah State University Press
An imprint of University Press of Colorado
245 Century Circle, Suite 202
Louisville, Colorado 80027

 The University Press of Colorado is a proud member of
the Association of University Presses.

The University Press of Colorado is a cooperative publishing enterprise supported, in part, by Adams State University, Colorado State University, Fort Lewis College, Metropolitan State University of Denver, Regis University, University of Colorado, University of Northern Colorado, Utah State University, and Western State Colorado University.

∞ This paper meets the requirements of the ANSI/NISO Z39.48-1992 (Permanence of Paper)

ISBN: 978-1-60732-679-3 (pbk)
ISBN: 978-1-60732-680-9 (ebook)
DOI: https://doi.org/10.7330/9781607326809

Library of Congress Cataloging-in-Publication Data
Names: Bridgeford, Tracy, 1960– editor.
Title: Teaching professional and technical communication : a practicum in a book / edited by Tracy Bridgeford.
Description: Logan : Utah State University Press, [2018] | Includes bibliographical references and index.
Identifiers: LCCN 2017025372| ISBN 9781607326793 (pbk.) | ISBN 9781607326809 (ebook)
Subjects: LCSH: Communication of technical information—Study and teaching. | Technical writing—Study and teaching.
Classification: LCC T10.5.T335 2018 | DDC 607.1—dc23
LC record available at https://lccn.loc.gov/2017025372

Cover illustration (c) Omelchenko/Shutterstock

# CONTENTS

# ACKNOWLEDGMENTS

A book is no easy task. Many people helped me along the way. From the very beginning, Kelli Cargile Cook contributed in immeasurable ways to my preliminary thinking and planning. Joan Latchaw and Kathy Radosta listened to me talk about it *ad nauseam* and provided valuable feedback on the introduction. Pavel Zemliansky and Karla Saari Kitalong also read the introduction, offering feedback from the field. Finally, Jonathan Santo helped me figure out how to conduct and read the results of a survey I conducted. And even though I didn't end up using the survey results, I learned much-appreciated lessons about quantitative research.

# TEACHING PROFESSIONAL AND TECHNICAL COMMUNICATION

# 1

## INTRODUCTION TO *TEACHING PROFESSIONAL AND TECHNICAL COMMUNICATION*
### *A Practicum in a Book*

Tracy Bridgeford

*Teaching Professional and Technical Communication: A Practicum in a Book* grew out of my efforts to create a technical communication pedagogy course for local secondary education teachers, part-time teachers, and graduate students who knew little to nothing about the subject, let alone how to teach it. This book delivers what I didn't have when I first taught technical communication—a practicum that enabled me to see pedagogical approaches in action before stepping into the classroom. This collection is intended to help inexperienced instructors understand the classroom experience of the PTC instructor and how to be professional and technical communication instructors in face-to-face classrooms. *Inexperienced instructors* refers to instructors from academia with no industry experience, industry professionals with no academic training, or graduate students with neither. To address this gap, I thought it was important to require readings of the landmark essays that provide a theoretical foundation informing pedagogical approaches (see Suggested Readings at the end of this introduction), but which also provide pragmatic knowledge about instruction. Because many of us in the field learned to teach Professional and Technical Communication (PTC) through trial and error, hallway conversations, conference presentations, and discussions with colleagues—all of which address important theoretical information about teaching professional and technical communication—many of the practical aspects of teaching the subject have not been available in print since the 1980s, and so much has changed since those early days.

Although it does not aim to be a compendium of best practices, this collection does provide plenty of practical advice and examples. To that

DOI: 10.7330/9781607326809.c001

end, I asked contributors to shape their chapters as if they were observers in a classroom recording classroom practice. They describe what teaching a particular PTC competency, such as information design, looks like in actual practice by establishing a scenario; providing a theoretical basis as a foundation for interpreting the scenario; illustrating the practical aspects of applying the approach, method, or practice; and describing assignments or activities that instructors can generalize to use in their own classrooms. Each chapter concludes with a list of questions for pedagogical discussions. It delivers a deeper level of training—a practicum that prepares instructors to walk into the professional and technical communication classroom with confidence. The term *practicum* can suggest a purely practical approach to teaching, or it can refer to the cumulative knowledge and skills acquired over the course of an education. For this collection, practicum signifies both the theoretical and practical aspects of preparing to teach PTC. This "practicum in a book" guides instructors through the teaching of topics normally covered in service or introductory courses in professional and technical communication.

I begin this practicum in a book by describing the problem-solving approach used most frequently in professional and technical communication classes followed by a description of the various competencies taught in these classes. Technical communication instructors must be aware of the role these competencies play in writing technical documents so they will be better able to guide student learning. These general competencies include audience analysis and purpose, information design, project and content management, style, and ethics. Although I discuss each competency separately, they are typically taught simultaneously. That is, it is difficult to teach ethics without also considering rhetorical devices such as audience and purpose or to teach genre without also addressing design and content strategy. Likewise, it is impossible to teach any of these competencies without also tending to style issues. And given the nature of globalization, it would be difficult to prepare technical documents for international contexts without also considering the impact of these competencies on those audiences. This problem-solving approach helps instructors situate these competencies within a context of social action.

## PROBLEM-SOLVING APPROACH

Sometime in the 1980s, we moved from a forms-based approach focused on the various parts of a form that students followed like a template with little consideration for the action involved to more socially based approaches that examine the contexts and influences on that docu-ment—what I'm calling a *problem-solving approach*. This approach is a critical thinking method that guides students through the various itera-tions of a technical document. It asks students to approach their writing from the standpoint of solving a communication problem. For example, while a memo as a form has identifiable, common parts (i.e., To, From, Date, and Subject), it is equally important to consider the various social aspects of that piece of communication and why, for example, this or that word, heading, or design was chosen. Documents grow out of a context and a situation, which affect all aspects of the writing. S*ocial aspects* refers to the various contexts in which PTC is involved, such as examining the power relations between the addressee and the writer, or the role of professional and technical communicators in an organiza-tion's hierarchy, or how the creation and organization of content (seen as a product) can help define those relationships. These examples dem-onstrate the value of focusing on the social approaches of PTC in ways that engage students in their own learning and help them develop an awareness of audience, purpose, and situation. Lloyd Bitzer (1968) calls this the "rhetorical situation."

The exigency of a rhetorical situation, Bitzer says, is what calls the writing "into existence" (its purpose or reason for existing) and what informs the writer's choices about the appropriate style, tone, register, design, and graphics to be used given a particular situation (2). When students learn about writing in a context, or reacting to a situation, they begin to see how communication happens in the workplace. It's not simply the creation of a genre; it is a form of social action that grows out of a particular situation. In the workplace, events often require a docu-ment of some sort (e.g., a trip report, an activity log, an instruction set, a user manual, a memo, a letter, and so on) that communicates various actions to a specific internal or an external audience (e.g., new policies are enacted, updates are communicated, marketing materials are made, and so on). Creating assignments that focus on a situation students can then use to direct their writing (e.g., writing a progress report for a group project) has become common pedagogical practice. Shaped by the nuances of the rhetorical situation, instruction has evolved into a pedagogy focused on problem solving, and this approach is what enables students to become agents writing in situ.

The various stages of this approach include preparatory work of such considerations of audience, purpose, research, genre, and situational analysis, that is, how the document will be used and in what context. For example, when I assign instruction sets, I often begin by showing

students images from NASA that show an astronaut making repairs outside the International Space Station. In this image, an astronaut is consulting the pages of a book attached to his wrist; the book is a big picture
book made out of plastic. In another image, an astronaut is consulting a
portable tablet attached to her wrist. Both images are powerful remind-
ers that documents (print or digital) created for space must address how
they will be used by the intended audience. Given the confines of space
travel, astronauts need large text and images that are easy to see despite
the huge helmet and big buttons that are easy to push with large gloves
on. This example illustrates the importance of social context and the
kinds of knowledge writers need when drafting documents that will be
used for a particular purpose.

Following this preparatory work, instructors can then instruct students
to begin drafting the document, using what they learned from the preparatory work to craft sentences and organize the document's content.
Students then must think about how best to organize the content in ways
that make it easy to find information because technical documents are
not often read from beginning to end as one would read a novel. They
are scanned by readers looking for specific information. For example,
astronauts repairing a loose joint on the International Space Station
may bypass some information about what the joint is and how it works
in order to get to the repairs more quickly. Given their limited time
outside the space station, astronauts must make repairs quickly and cannot spend time on information not useful to the specific task. Students
must then think about the wording of each step, which requires using
the imperative voice, providing feedback statements when necessary,
shaping the content into manageable chunks. Following the drafting
and organization stages, students would complete the writing process
by copyediting and proofreading the content, paying close attention to
sentence and paragraph structure and style.

Students would then ensure that the design of the document aids
usability. By discussing this step as if it were a last part of the process, I
do not mean to suggest that design instruction is saved for the end of
the process as if it is an afterthought or purely decoration. Design issues
are raised throughout the process and are certainly a part of all steps in
the problem-solving approach. Design issues involve the presentation of
content for a specific audience's use. As mentioned earlier, astronauts
working outside the space station need large pictures and text to work
effectively. Instructions must be designed with one step per page/screen
to accommodate a larger font or image size as well as huge gloves. In this
way, writers must provide comprehensive information in each step, all

while being succinct, so an astronaut would not have to continually page/move back and forth from step to step. The problem-solving pedagogical approach I just described is evident in each chapter of this book no matter what competency the PTC authors discuss in the pages to follow.

## Audience Analysis and Purpose

Audience analysis is the primary competency PTC students must engage in if they are to become effective communicators on the job because it influences every other aspect of technical documentation, such as style, tone, organization, and design. In my own and many other instructors' experiences, students tend to skip the necessary audience-analysis work, mostly because it involves changing from an "I" to a "you" attitude, as both Jim Dubinsky (chapter 2) and Dan Jones (chapter 3) describe in their respective chapters. Switching viewpoints is challenging for students because past experiences in composition, for example, have shown them that prewriting activities such as invention heuristics, critiquing both strong and weak writing, and instructor comments on drafts are the means for getting started on writing or on designing projects. Once students understand the value of the "exigency of audience awareness," as Tharon Howard calls it in chapter 11, the more effective their documents or presentations will be. As a central tenet, audience analysis is the most important work of writing a technical document in that it is what lends credibility to the writing. When a document is well written, the work of the user can continue without interruption and without questions to a call center. The credibility of a well-written document is especially evident when translating documents for international audiences because understanding the culture of a communication problem, whether familiar or unfamiliar, is a necessary step toward writing and designing technical documents.

## Information Design

Creating technical documents involves various aspects of the design process, including genre, visual cues, graphics, and information design. Part of solving the problem of the design of a document is conducting a genre analysis, choosing the appropriate genre, and shaping its design in ways that address the audience and situation. As a starting point, we might ask, What does the genre look like? How will it be used? What social action is it addressing? We might look at models, but we must be prepared to adjust the design as needed based on what we've learned from other analyses as well as the situational requirements. Fundamentally, genre is a social, rather than an individual, process, a process that can help instructors fight the ivory-tower conception of writing most students harbor.

Design also involves examining how people use text and images to provide visual cues about a document's structure and organization, such as text, pictures, italics, boldface, type size, white space, and positioning of elements on a page—all of which should make it easier for the reader to find the needed information. These symbolic aspects of a document are important because technical documents are scanned, not read in their entirety. For example, when assigned a proposal or report on the effects of global warming, students must understand that different people will read the document in different ways. The executive will probably only read the executive summary, while a financial agent may only be interested in the budget. As such, the various areas of the report must use visual cues such as bolded headings or bulleted lists to designate different parts and use white space effectively in order to highlight important information such as facts and figures or graphics. Design is even more important when preparing documents for translation because credibility is an important factor: design expectations may differ depending on the culture. It is important to know, for example, what colors or images are appropriate to use. Design has become more centralized in PTC pedagogy because readers have become more visually oriented.

Design should not be separated from any other competencies when writing technical documents because it is as much a part of documentation as writing, so it's important to emphasize its rhetorical nature. For students, this competency may seem largely minor at first, but it is every bit as rhetorical as writing. The design of a document is not merely applying decorative characteristics as a finishing touch. Design should provide instruction to the user in ways that demonstrate how to read the document. In this way, it compels the user to act in a certain way and, therefore, should be emphasized throughout the course, not just in a special chapter or unit. Design decisions can mean the difference between the text being taken seriously or ignored. For example, choosing a font indicates the type of information presented. When the Higgs boson particle discovery announcement was made, the most significant scientific advance in forty years, it was marred by the use of the font Comic Sans, a childlike font typically used in informal situations. Using it to announce a formal, major scientific advancement gave the impression of frivolity and not the seriousness the situation called for. Design competency is addressed in both service and major's courses because technical documents are not just written; they are designed to ensure readability, legibility, and usability. Readability refers to how easy it is for users to find the information they need, while legibility refers to a font's appearance and its ability to be deciphered. Usability refers to how well the document can be used to find what's needed. Everything about a document, or information design, is about how content is presented to an audience.

### Content and Project Management

Whether the task is content management, or project management, students most assuredly spend a significant time managing their tasks, which Dave Clark in chapter 4 calls *content strategy*. Content strategy is a "movement," he says, that emphasizes single sourcing of content in ways that enable it to be used more than once and in more than one context. This management component then becomes a necessary part of work in the twenty-first century as workers connect large masses of information among various departments, all of which contribute to the overall content created for technical products. This information is "repeatable" in that it shapes information into repeatable blocks of content. Key to content strategy and its management is the use of "modular chunks" in topic areas that ensures they can be used in a variety of contexts, especially content written for translation. As a component of the problem-solving approach, content strategy involves completing a needs assessment, content inventory, and content audit—all of which involve situation, audience, and design analyses. Planning in these areas keeps students focused on writing as solving a problem.

### Ethics and Style

Two competencies integral to the first three previously described—audience analysis, design, and management—are ethics and style. Ethics plays a role in all situations, including power relations, organizational structures, credibility issues, stylistic choices, content-management strategies, and genre choices. Sometimes ethical dilemmas can seem small and insignificant (e.g., stealing a pen from an employer) while other issues are major, affecting many people (e.g., ignoring warnings for faulty O-rings, such as with the *Challenger* disaster). Some instructors discuss ethics in a specialized unit while others include ethical discussions throughout the course as a part of each assignment. Because ethics cannot be separated from the rhetorical situation, engaging in its discussion throughout the term is pedagogically responsible.

I recommend that instructors research a local situation, bringing into class real documents and asking students to analyze them for ethical considerations. In 2002, while a graduate student at Michigan Tech, I used a novel about a global environment issue in conjunction with a case study about Torch Lake in the Upper Peninsula of Michigan in order to engage students in an ethical discussion involving a memo they were writing to an environmental-agency supervisor. Specifically, I have used Scott Russell Sanders's novel *Terrarium* as a context for assignments because it tells a dystopian story of a global environmental crisis that has forced humans to move inside gigantic, domed enclosures in order to protect them from the toxins of the earth. I used the Torch Lake case study alongside the novel because it had been designated by the Environmental Protection Agency as an Area of Concern due to its deteriorating water quality resulting from all the copper mining in that area. In preparation for an assignment, I asked students to discuss the ethics involved in both narratives. One student wrote that "Torch Lake is the exigency of *Terrarium*," meaning that if humans continue to pollute the water as demonstrated in the Torch Lake case study, they could end up in an enclosure as depicted in *Terrarium*.

Instructors might use the example memo in Stephen Katz's (1992) article "The Ethics of Expediency: Classical Rhetoric, Technology, and the Holocaust," or the memos provided in Carl G. Herndl, Barbara A. Fennell, and Carolyn Miller's chapter "Understanding Failures of Organizational Discourse" in *Textual Dynamics of the Professions: Historical & Contemporary Studies of Writing in Professional Communities* (Herndl, Fennell, and Miller 1991). As Katz says in his article, the memo he discusses, for all intents and purposes, is a good example of a well-written memo. However, it is ethically corrupt in that the memo is talking about human beings on one of the trains en route to concentration camps using a euphemism ("merchandise") to refer to them. Herndl, Fennel, and Miller analyze memos concerning the accident at Three Mile Island and the shuttle *Challenger* disaster. In both cases, they found that the accidents happened due to "misunderstanding and miscommunication" (279). One of the memos about the *Challenger* disaster and the faulty O-rings, for example, was not taken seriously because the author, a budget analyst at NASA, had only been with NASA for a few weeks. He was seen as a newbie without the proper training and knowledge because he lacked detailed data, quantified budget estimates, and subheads; used nontechnical language; and mentioned safety concerns in a budget memo, so he was regarded as an "outsider" (299–300). His memo was therefore dismissed. These memos are just a few examples that have proven successful as examples of ethics in the technical communication classroom.

Another competency distinctly a part of audience, design, and management analysis is an embedded competency that is always present even when no one mentions it: style. Style refers to the choices we make when writing and designing a document. In his book *Technical Writing Style*, Dan Jones's (1998) definition of style can help us contextualize this competency: "Style is your choices of words, phrases, clauses, and sentences, and how you connect these sentences. Style is the unity and coherence of your paragraphs and larger segments. Style is your tone—your attitudes toward your subject, your audience, and yourself—in what you write" (Jones 1998, 3). Although Dan Jones also discusses style comprehensively for this collection, each author in this collection directly or indirectly addresses style issues because it is impossible to engage in any other competency without also addressing the stylistic issues that affect meaning, such as comma placement and sentence-level changes. Because technical documents almost always start with a definition, I like to begin the content of the course by asking students to write definitions. For example, using *Terrarium*, I ask students to write a definition of a term used in the novel. Students sometimes invent terminology within the context of the story, such as, "Retina Scan Sign is a device confirming comprehension of the Mating Ritual statues." Weaknesses in this definition include using vague language (*device*) and illogical connections (*how can you measure comprehension with a retina scan?*). Another student wrote, "The Intragaming System is a system in which players compete within a community." This definition is repetitive (*system*) and vague (*a community*). However, in the following example, a student wrote, "A belt transect quadrant is a measured and marked rectangle used in ecological studies to divide a larger section of land into smaller, equally sized subparts."This definition is more effective in that it properly identifies the type (*measured and marked rectangle*) and its difference from other belts (*used in ecological studies*) and uses exact language (*equally sized subparts*). Style, then, is a matter of making decisions about language use.

Style and ethics, as I see them, are issues of credibility. Credibility is, as Kirk St.Amant says in chapter 13, "what drives us" to present documents in ways that accurately address the culture of the audience, which is what determines what is and is not credible. If instructions are not clear, users will not use them. If information is buried, readers may miss important content. From both a style and an ethics perspective, if the design interferes with the reading of a document, users will ignore the document.

If an information graphic is duplicitous, the content might be regarded suspiciously. If a website shows bias toward a particular culture, users will click away from it. For example, the O-ring warnings to the manufacturers of the shuttle *Challenger* were presented in a text-heavy PowerPoint slide, buried within other, more informative information—as opposed to emphasizing this warning. It was easily missed or ignored. Attention to style always involves making ethical choices—major and small—and how these decisions are made determines the level of the document's (print or digital) credibility.

## CHAPTER SUMMARIES

Instructors teaching technical communication tend to teach what's referred to as the *service course*, a course that mostly serves nonmajors. Because PTC service courses often constitute the required third writing course for STEM and other nonmajors such as criminal justice or aviation, emphasis on the situational and contextual aspects of writing in the workplace is, indeed, an important and appropriate stance to take in the classroom. Students who take PTC service courses, and even majors taking introductory PTC courses, have little to no knowledge about workplace writing and the dynamic contexts in which it occurs. They may be familiar with some workplace genres but will need convincing that audience and situational analyses are important and necessary. Altogether, the chapters in this collection constitute a term's course in action.

The chapters probably work best when combined; that is, you could ask students to write and design documents for an international audience and discuss the appropriate style and tone as well as the appropriateness of their designs based on the culture. You could ask students to write a set of instructions and engage in usability testing with other students on campus or analyze and create information graphics, all the while discussing ethical ramifications of word choice and instructional design. You might ask students to work on collaborative projects that require them to research an historical context such as Dombrowski's cigarette ads or Dubinsky's *Challenger* memos and discuss the cultural influences on language use. My own approach grows out of communities of practice theory; I ask students to read and interpret a narrative context (e.g., *Terrarium*) for assignments in ways that engage the social and ethical considerations of a particular culture, as I mentioned in an earlier section (see Bridgeford 2007 for further description). Any one of these assignments/approaches could then require students to present their information in class using slides. Although I recommend starting with chapter 2 given its rhetorical focus on audience and situation, the chapters that follow do not necessarily need to be read in the order in which they appear. The chapter summaries that follow can help you decide where you want to start.

In the second chapter, James Dubinsky appropriately focuses on the rhetorical situation (audience, purpose, and situation). This rhetorical foundation is central for understanding basic concepts of technical communication. Through that lens, he models the rhetorical situation's influence through the presentation of three different analyses. Using the *Challenger* disaster as an example, Dubinsky examines three memos associated with the O-ring design and how persuasion plays a role in the writer's language use, the document's design, and the word choice. He then compares two position descriptions, demonstrating the importance of genre analysis and disciplinary knowledge. Overall, Dubinsky argues that the "disposition" of the rhetor is vital to the rhetorical situation in that it heavily influences the writing in important ways.

Style, the focus of Dan Jones in chapter 3, stresses the importance of teaching students about clarity, conciseness, courteousness, and persuasion in technical communication—all of which "make a difference in their success on the job." Jones's comprehensive approach, which can involve possible role-playing activities and assignments, begins with the essentials of style such as audience expectations, usage, and word choice. Emphasizing more advanced concerns, Jones then explores the "concept of discourse communities," which can help students understand the writing challenges they will face in their fields. Toward the end of his chapter, Jones argues for a more "advanced level" that enables students to explore the concept of discourse communities.

In chapter 4, Dave Clark begins with the task of defining *content strategy*, a term still in flux. Generally, Clark explains, content strategy is a combination of content delivery, content acquisition, content management, and content engagement—all characteristics of a process that involves connecting content to overall organizational processes and planning. Clark draws from various valuable resources available from practitioners to describe this relatively new approach. Technical communication, Clark concludes, can no longer take the "clear, singular path it once" did; it now involves an applied literature review, needs assessment, content inventory, content audit, and tool training—all important aspects of teaching content strategy.

Brent Henze, in chapter 5, differentiates between the "traditional approach" to teaching genre (i.e., introducing models with specific rules for layout and content) and the more current approach that emphasizes genres as "social action." It's more pedagogically sound, he explains, to view genres from the theoretical standpoint that they are responses to social situations. In this way, he says, students learn to respond to writing situations in ways that require them to work out the appropriate genre and its characteristics for a specific situation. Henze sees genre theory and teaching genre as a way of "conceptualizing writing" as part of social context that requires specific responses. His approach is effective for teaching students the difference between formatting a particular genre and writing, designing, and creating an appropriate genre. "All writing," Henze says, "responds to prior communication," enacting a rhetorical situation.

Karla Saari Kitalong focuses chapter 6 on the state of information graphics, which has not received as much attention as it should have in pedagogical discussions. She then describes Stuart Selber's (2004) model of technological literacies (functional, rhetorical, and critical) and how they work together to help students develop more quantitative competencies. In this examination, Kitalong compares three textbooks' treatments, concluding that these textbooks' treatments "illustrate that their content has remained much the same" since 2007. Kitalong argues that it is important that "ethics . . . not be ignored in discussions about information graphics," which the major textbooks rarely do, often pointing to professional obligations and codes of conduct as constituting ethical discussions. She concludes by describing her pedagogical strategy for teaching information graphics using Selber's "multiliteracies approach" and the importance of critical-literacy practices with information graphics that require attention to citations to avoid "reader deception."

In chapter 7, Eva Brumberger encourages instructors to incorporate design principles into all aspects of a technical communication course as opposed to focusing on design only for a single unit or chapter. She sees design as a problem-solving process that affects all rhetorical aspects of a document because design "shapes users' interactions with an information product." She begins with a review of Gestalt principles that provide the foundation for design pedagogy and offers some activities and assignments for integrating design into courses. Throughout the chapter, she notes some useful resources for expanding instructors' knowledge about design. Designing a document, whether in print or digital mode, is a matter of, she says, "solving a communication problem."

In chapter 8, "Designing and Writing Procedures," David K. Farkas explains the fundamentals of procedures, focusing on the domain, system state, audience, and medium and modalities, key concepts for writing clear and concise how-to discourse. Using a fictitious teacher,

Farkas explains the best way to go about teaching each part of a procedure, offering pedagogical advice and exercises and assignments, which
include attention to various modalities (e.g., paper, video, and audio).
He concludes with advice on how to include all or parts of his discussion
just in case time and length of class are factors. Farkas's chapter takes
you through the writing of procedures using clear and specific examples
that demonstrate effective instructional techniques. Farkas emphasizes
that when writing procedures, there are many considerations of what
types of visuals to use.

Emphasizing the importance of teaching ethics in technical communication classes, Paul Dombrowski in chapter 9 emphasizes that
ethics should not be reducible to a single chapter or unit. He believes,
and rightly so, that an expanded view is necessary. He offers a "primer"
that can be weaved throughout a course, presenting three modules that
emphasize important documents associated with a smoking advertisement, global climate change, and the shuttles *Challenger* and *Columbia.*
His chapter encourages us to be "mindful" of the ethical dilemmas
present in our field. Dombrowski addresses stylistic issues when he demonstrates his pedagogical method of rhetorical analysis and ethics that
includes important perspectives such as the "ethical care of the writer"
(to make clear what is being communicated) as well as the "ethical
responsibility of readers" (to read and understand). This comprehensive treatment of ethics is vital to all other competencies.

In chapter 10, Peter S. England and Pam Estes Brewer explore the
meaning of collaboration and the impact on learning that occurs when
individuals work together. They argue that collaboration helps students
become more engaged and that the learning that occurs through collaboration reflects workplace contexts. Their chapter recommends
instructional methods based on the tenets of instructional design as well
as technical communication education. They also include recommendations for using, managing, and assessing collaboration in the technical
communication classroom. Through carefully crafted examples from
each of their own classes, England and Brewer provide vivid observations of teaching collaboration in action.

While taking the reader through the basics of usability testing in
chapter 11, Tharon Howard argues for the need to focus "explicitly on
integrating user-centered design processes" by incorporating usability
testing into assignments. He takes the reader through five phases of
usability testing, which include (1) establishing research questions,
(2) planning, (3) collecting data, (4) analyzing and coding data, and
(5) writing a recommendation report. Usability testing, he says, is a

"mainstream skillset" necessary for students in technical communication classes. He identifies useful guides, both in print and online, for learning more about usability testing, fulfilling a "credibility gap" that helps technical communicators develop confidence not only in the classroom but also in the workplace.

In chapter 12, Traci Nathans-Kelly and Christine G. Nicometo describe presentation techniques for use by instructors of STEM students, instructional approaches for convincing students (and peers) that these techniques are valid, and an alternative method to bullet-heavy presentation slides. Along with examples of these adaptable methods, the authors describe hand placement, laser pointers, and eye scanning. The last part of their chapter addresses agile best practices for slide use that involve using complete sentence headers, bigger and better visuals, and speaking notes. They offer sound advice through engaging examples, encouraging instructors to teach students to "make the most of slide acreage" by using one complete sentence and one image per slide and by using the notes section for all of the text. I've adopted this practice in my own and students' presentations and have experienced success in doing so.

In chapter 13, Kirk St.Amant focuses on international and intercultural communication, an important aspect for today's technical communication courses because students, whether they plan to work as technical communicators, engineers, or in some other job, will likely be creating artifacts "for individuals from other cultures." He describes a pedagogical approach he calls "comparative online analysis of cultures (COAC)" that uses heuristics to help instructors focus students' attention on two factors specific to international communication: credibility and usability. He offers advice on how COAC works in the classroom and demonstrates it with various assignments. His pedagogy can be used in any technical communication course. He argues that "culture and credibility are (inter)connected" in that the higher the credibility, the more likely readers/users are to pay attention; "It drives us," he says. Even more important is his point that credibility is not inherently a part of any document (print or digital). It is taught through culture, which determines what is and is not credible.

## Suggested Readings

If you are interested in learning more about the history and practice of technical communication pedagogy, I recommend you start with the following two collections:

Johnson-Eilola, Johndan, and Stuart Selber, eds. 2004. *Central Works in Technical Communication*. New York: Oxford University Press.

Johnson-Eilola, Johndan, and Stuart Selber, eds. 2013. *Solving Problems in Technical Communication*. Chicago, IL: University of Chicago Press.

## References

Bitzer, Lloyd F. 1968. "The Rhetorical Situation." *Philosophy and Rhetoric* 1 (1): 1–15.

Bridgeford, Tracy. 2007. "Communities of Practice: The Shop Floor of Human Capital." In *Resources in Technical Communication: Outcomes and Approaches*, edited by Cynthia L. Selfe. Amityville, NY: Baywood.

Herndl, Carl G., Barbara A. Fennell, and Miller, Carolyn. 1991. "Understanding Failures in Organizational Discourse: The Accident at Three Mile Island and the Shuttle Challenger Disaster." In *Textual Dynamics in the Professions: Historical and Contemporary Studies of Writing in Professional Communities*, edited by Charles Bazerman and James Paradis. Madison: University of Wisconsin Press.

Jones, Dan. 1998. *Technical Writing Style*. Boston: Allyn and Bacon.

Katz, Stephen. 1992. "The Ethics of Expediency: Classical Rhetoric, Technology, and the Holocaust." *College English* 54 (3): 255–75.

## 2

# RHETORICAL ANALYSIS
## A Foundational Skill for PTC Teachers

James M. Dubinsky

The list of chapters in this book demonstrates the breadth of our field of professional and technical communication (PTC). Chapters on definitions and instructions are presented alongside ones on content management and information design. Some of these chapters have been included in textbooks for generations; some are relatively new. But all have at least one thing in common: they ask you to teach students *how to use* symbolic language in specific contexts. This focus on *how* people *use* symbols, whether the symbols are gestures, pictures, or words on a page or screen, to influence perceptions, beliefs, behavior, or even action or activities is a focus on *rhetoric*. A presupposition of rhetoric is that a rhetorical situation exists involving an audience, an exigence relevant to both the audience and the speaker/writer/designer, and constraints.

Beginning your semester with a focus on rhetoric and rhetorical analysis may seem odd or might appear to be a tough "sell," particularly to students coming from engineering or science where the term *rhetoric*, if used, often has the pejorative or diminutive connotation of "mere rhetoric."[1] Although many students have had some introductory writing course in composition, much of that knowledge has been pushed way back in their memories as they have struggled to learn the principles and application of those principles in their chosen fields of mechanical engineering, biochemistry, or plant pathology (to name a few of the disciplines that often require PTC courses). For that reason, a number of beginning teachers shy away from rhetoric and choose what seems to be an easier path: immediately building a bridge to the students' disciplinary knowledge—writing instructions, creating reports, drafting interoffice communication.

I want to demonstrate why beginning with a focus on disciplinary knowledge and disciplinary texts, without providing key foundational knowledge that transcends disciplines, should be the road *less traveled.*

DOI: 10.7330/9781607326809.c002

For millennia, rhetoric has been taught as central to communication in and across disciplines (Joseph 2002). Scholars, teachers, and writers in a wide range of disciplines rely on it as a means of "explaining how texts work by connecting rhetorical strategies to their effects on historical audiences" (Ceccarelli 2001, 6). Those scholars explain that their findings "are written with particular audiences in mind and reflect the presumptions carried by authors regarding the attitudes, expectations, and backgrounds of their intended readers" (van Maanen 2011, 25).

Rhetoric is a way of using knowledge to create understanding and shape opinion. As Walter Ong has said, "Once writing made feasible the codification of knowledge and of skills, oral performance [rhetoric], enjoying the high prestige that it did, was one of the first things scientifically codified" (Ong 1971, 4). Rather than mere frippery, rhetoric is an art with centuries of thought and use behind it. Using writers such as those I've just cited (Leah Ceccarelli, who focuses on understanding and "shaping" science, or van Manen (1994), whose work crosses boundaries among business, law, criminal justice, and management) demonstrates the usefulness and value of beginning PTC courses with a foundation in rhetoric and rhetorical analysis. Focusing on those historical audiences and intended readers, as well as on the reasons for writing (purposes) and the situations the writers find themselves in (arguing, explaining, outlining), helps your students not only communicate more effectively within their disciplines but, perhaps more important, also helps them communicate what they *know* to others who lack their specialized knowledge.

My intention, then, is to provide theory and praxis: an understanding of why you should teach rhetorical analysis and practical strategies (e.g., assignments and activities) for doing so. My plan is to provide some background material and then walk through two rhetorical analyses, focusing on memos and job advertisements, two ubiquitous genres that are familiar to all students. I rely on a document that may seem a bit odd or old: a 1979 memo from an engineer working on the space shuttle *Challenger* prior to its 1986 launch and subsequent explosion. I choose it for its historical and cultural significance and its relevance to both intra- and interdisciplinary communication.

Virtually every student knows about the space shuttle *Challenger*. Even though a previous fire on the launch pad had killed several astronauts,[2] NASA had also enjoyed the Apollo successes, including the moon landing. The space shuttle program was intended, by President Reagan, to become "ordinary." Space shuttles were beginning to be compared to trucks. Because they were becoming so ordinary and familiar, NASA needed to take what seemed to be a spectacular step: ask Christa

McAulliffe, an elementary schoolteacher, to join the crew of astronauts. What makes this memo and the situation so much more spectacular and tragic was that the entire nation, to include Christa's students, watched the launch and subsequent explosion. The vivid images of the *Challenger* blowing up and falling out of the sky in pieces were replayed over and over. Much like the JFK assassination, most people alive then could tell you exactly where they were when they heard or learned about the disaster. Much like the 9/11 tragedy, the footage of the launch and explosion, seventy-three seconds after liftoff, was literally seared into our country's psyche.

Equally relevant is the fact that the subsequent investigation found a number of potential reasons for the accident, but in the end the technical reason was the O-rings, which are the subjects of the memo. O-rings and engineers, and the situation facing those engineers as they tried to explain the mechanics of the O-rings to managers, create a context, a rhetorical situation, that fits well with PTC students.

PTC students are able to relate to Mr. Ray, the engineer, and they find his memo accessible. Students quickly understand and think about ways to apply the rhetorical lessons they gain from this analysis. You can use this memo and then lead into discussions of memos or letters from more recent situations. As for the lessons to be learned from the job ads, they are fairly easy to see: job ads are a genre students must learn how to *read and understand* to have the best possible chance of getting an interview.

As a result of these two extended exercises, you will be able to ensure that students know what twelfth- and thirteenth-century theorists realized: whenever writers write (or speakers speak or designers design), they do so *in some way* hoping to achieve *some goal*. Learning to identify those ways and then learning strategies for audience and situational analysis are foundational requirements for students of professional and technical communication.

## A BIT OF HISTORY: RHETORICAL ANALYSIS
## AND WRITING STUDIES

In 1874, the Rev. J. McIlavine published "Rhetorical Analysis and Synthesis" in the *Presbyterian Quarterly and Princeton Review*. The opening claim is powerful: "Analysis and synthesis are the most fundamental processes of discursive thinking and expression" (456). McIlavine argues these processes are the "diastole and systole of the mind" that provide life-giving force to "quicken and nourish" mental activity (456); he then says these two processes "must be mastered . . . [and are] indispensable to freedom and power in all the subsequent processes of rhetoric" (457). Admittedly, Rev. McIlavine's focus was neither on undergraduate education, broadly writ, nor on the teaching of professional and technical writing. Rather it was on preparing ministers to serve their congregations, ensuring that they could serve best by making relevant issues come "alive" during their sermons. He was teaching a way of thinking, which according to many, including Howard Gardner[3] (Booth 2004, 98), is the most important skill.

McIlavine's emphasis on the foundational importance of teaching rhetorical analysis as a way of thinking remains with us today. Nearly 150 years later, Martin Medhurst (1998) explains that "rhetoric does not operate apart from the world in which we live," and "the coursework which brings it alive in all its variety is crucial to the undergraduate experience" (347). I concur and intend to demonstrate that rhetoric and the process of rhetorical analysis not only teach "us how to think; [they help] us understand how we create, maintain, and change ourselves and our roles in the world; how we fashion and modify our societies; how we approach others in this increasingly complex world of ours" (Turner 2013, 188). Teaching rhetorical analysis then becomes important at micro- and macroeducational levels. Students learn to *see* texts differently: as technologies for "allowing responses between various audience attributes" (Mehlenbacher 2013, 195); as technologies with genre characteristics (Miller 1984; Russell 1997); as part of larger, organizational systems or cultures (Spinuzzi 2004; Yates and Orlikowski 2002).

I have argued elsewhere about the importance of integrating a rhetorical approach into teaching: "Our work involves more than teaching our students strategies or forms; it also involves asking them to consider the impact of those strategies and forms on public policy. We teach them to become user-centered practitioners, to take their audience and its needs into consideration always" (Dubinsky 2004, 4–5). In this earlier essay, I cite Dunne's argument that "the main aim of 'educational studies' should be to contribute to the development of this disposition" (1993, 369), which can be explained as a way of coming to knowledge through reasoning that is the result of their own analysis.

## INSTILLING A DISPOSITION: MOVING FROM IGNORANCE TO KNOWLEDGE

Instilling a disposition in students as a result of teaching rhetorical analysis in any course in a professional writing curriculum is perhaps the most important outcome we can hope to achieve. Most students who take professional writing classes, particularly those taking our service classes, do not have a lot of experience writing in professional settings. They are not familiar with the genres of practice in the workplace; they do not have a keen understanding of the "messy little facts, opinions, beliefs, motives, documents, texts, personalities, circumstances" (Medhurst 1998, 347) that make the world of work run. We help them understand frames such as *contexts* and *actors*. We help them understand the *relationships* and *patterns of interaction* (Spinuzzi 2013, 263–65).

I have found, like many others before me (Anderson 2014; Crane 1956; Markel 2014),[4] that models are useful resources. In his *Handbook of Rhetorical Analysis*, John Franklin Genung (1888), then a professor at Amherst College, wrote that he believed students needed to study "models of excellence" in order to understand what the "making of good literature involves" (1). Genung was following in the steps of rhetoricians such as Aristotle, Quintilian, and Cicero, who argued that "oratorical skills are acquired by three means—theory, imitation, and practice" (Corbett 1971, 243). Using models can introduce students to the "messy little facts," provide a foundation of knowledge and practice about the rhetorical situation, and introduce the idea that rhetoric is a form of social action (Hauser 2002; McKerrow 2013).

## Bringing It to Life

In nearly every course I teach, from the standard service courses in business or technical writing to graduate courses in pedagogy, I use a document in the first week of class that demonstrates how "rhetorical analysis models a way of thinking about discourse as it affects everyday life" (Turner 2013, 188): a memo from a midlevel engineer that was just one of many documents included in the *Report to the President* by the Presidential Commission on the Space Shuttle Accident (1986). As a first-year PhD student, fresh out of the military, I used this document in my first scholarly presentation (1993), and I included it as a part of an opening exercise in a locally published online textbook that two colleagues and I created in 2001 (Armstrong, Dubinsky, and Paretti 2001). The memo not only highlights the value of "communication as a process of social influence" (Turner 2013, 179), it also demonstrates the value of arrangement, style, and document design (Schriver 2013) and enables me to highlight ways in which form and function are intertwined.

A common question when conducting or performing rhetorical analysis is whether one should follow a method or formula. While there are sets of questions I provide students,[5] usually I do not follow a formula or systematic process. I agree with Richard McKerrow, who explains that "formulaic approaches have their value . . . but beyond the initial role, they quickly lose force" (McKerrow 2013, 114). Instead, I use a general set of questions similar to those outlined by Medhurst, questions that focus on such issues as purpose, audience, and style (2013).

### The Rhetorical Situation

Regardless of the questions, I recommend starting with the larger, or macro, issues of communication to frame the social and historical context. For this situation, one strategy is to ask whether any students know when the space shuttle *Challenger* blew up. Most don't know it blew up in 1986, but some know why it did—usually an engineering student, who mentions something about "O-rings." If I'm lucky to have such a student, I congratulate them, and then I tell the class we will look at an official NASA memorandum that focuses on problems with O-rings in the rocket motor, problems that ultimately did lead to the catastrophic disaster of NASA mission 51-L, the destruction of the space shuttle *Challenger*, and the death of all seven crew members. I share the findings from the *Report to the President* (Presidential Commission 1986) that included the statement, "The loss of the Space Shuttle Challenger was caused by a failure in the joint between the lower segments of the right Solid Rocket Motor. The specific failure was the destruction of the seals that are intended to prevent hot gases from leaking" (40).

What I found fascinating, and a fact that surprises nearly every student, is that engineers at NASA knew there were problems with the seals long before the *Challenger* exploded. In fact, they recognized there was a design flaw in the solid rocket motor as early as 1977 (Presidential Commission 1986, 123). This design flaw was pointed out in two memos (one in January of 1978 and a second in January of 1979) that "strenuously objected to Thiokol's [the contractor for the rocket] joint seal design" (123). Both were written by Leon Ray, a Marshall (NASA) engineer involved with the solid rocket motors, and signed by John Q. Miller, chief of the Solid Rocket Motor Branch of NASA.

I then share a third memorandum—the sample text. I explain that in February of 1979, the same engineer, Leon Ray, made two visits to the manufacturers of the seals used in the joint between the lower segments of the right solid rocket motor—the joint mentioned earlier that failed catastrophically. Ray's memo describes those visits (see app. 2.A). During the commission hearings on May 2, 1986, Mr. Ray was asked about that memo. He could not recall hearing that any action had ever occurred as a result of the memo, and, in fact, he said that the "records show that Thiokol was [never] informed of the visits, and the O-ring design was not changed" (Presidential Commission 1986, 124; Covault 1986). Sharing this critical contextual information with my students illustrates rhetorical exigence as we keep in mind the "multiple considerations of audience and purpose" (Bernhardt 1986, 71).

### Starting with the Purpose

After the initial discussion, which can take anywhere from fifteen to twenty minutes, I ask students to read Ray's memo and comment on it in the broadest terms (e.g., do they believe it is well written?). A typical response focuses on the audience: considering that Ray was writing to an audience involved in time-critical actions with national importance, they believe the entire document seems wrong.

I then ask students to focus on what they believe is the critical information in the memo. The goal, obviously, is to *determine the purpose*. In virtually every class, students argue that the essential information is that both manufacturers of the seal thought it wasn't functioning as intended, a fact the designer of the motor—Morton Thiokol—needed to know early in the development of the shuttle. In the discussion that follows, we ponder why information they deem *critical* is "buried" within the body of the memo and why the memo reports mundane details of the trips as well as the rhetorical value of "mundane" discourse (Cyphert 2010, 351–52).

When I use this document, I focus on a wide range of issues, to include the importance of genre expectations and document design. In terms of genre, the cues point to the visits as the focus of the document. The subject of the memo indicates that this document merely describes the visits. The first sentence, *the purpose sentence*, claims exactly that. Each of the two other paragraphs, both rather long and involved and containing a tremendous amount of information, begin with the words, "The visit to." These genre cues point to the fact that the document is what is called a *trip report*. I ask them to do a quick Internet search for this genre, and they come back with several online resources such as the *Mayfield Handbook of Technical and Scientific Writing* (2001),[6] all of which mention trip reports should include "the reason for the trip, what was found, and one or more conclusions." Even as a trip report, it appears the memo should have had more impact, but when we put the document back into a work context, we discuss how a supervisor reading or scanning this memorandum might have missed the details about the O-rings, initialed it, and sent it to be filed.

## The Rhetorical Canons

At this point, we begin with a discussion of neo-Aristotelian strategies for analysis, focusing on the canons of organization, style, and delivery. Rather than walk through all the ways this document can be used, I'll focus on just one of the canons here: delivery.

Given the fact that the students are virtually all in agreement that the document failed, even as a trip report, I ask them to revise it, focusing on simple strategies for organization, style, and delivery. In the discussion of delivery, I mention visual cues (e.g., italics, boldface, different type size, white space, or the position of information on the page). The students recognize immediately that Mr. Ray did not use any such cues, creating a document that would require a reader *to actually read it in its entirety* to understand and process it. They agree that such a design seems ill-suited to the fast-paced world of NASA, or any business for that matter.

Once they've revised the document, usually in teams of two or three students, we come back together. Typically they revise with heavy emphasis on the critical information about the O-rings and almost no mention of the trips themselves. We share drafts and discuss the positive features, but then we have to return to the larger, organizational issues, what we could call the *macrocommunication issues*, sometimes described as ways to unite text, discursive practices, and social context (Huckin 1997). These issues bring the "macroanalysis of social formations, institutions, and power relations that . . . texts index and construct" (Luke 2002, 100) to light.

We discuss Mr. Ray's position as a subordinate, and I again mention his previous attempts to highlight the seal problem. Knowing his audience (the chief engineer of his branch—Mr. Eudy; the director of the Structures and Propulsion Laboratory—Mr. McCool; and the project managers for the motors at Marshall—Mr. Hardy and Mr. Rice), he would more than likely have written a straightforward, factual account rather than a sensationalized one. He would, however, have wanted to take the opportunity to highlight information that confirmed the data he'd twice raised a red flag about before.

Often the discussion starts to wane here, but some students ask whether a revised, more direct, more accessible memorandum would have received more attention. We conclude that there is no way to know the answer to that question. Expanding the discussion further, beyond the relationship of Mr. Ray to Mr. Eudy, I offer some additional information about other problems contributing to the disastrous decision to launch the *Challenger*. Much has been written about the lack of communication and the management problems (Arnold and Malley 1988; Dombrowski 1992; Herndel, Miller, and Fennell 1991; Romzek and

Dubnick 1987; Rowland 1986; Winsor 1988; Boisjoly et al. 1989). The problems the commission found did indeed exist. It is heartbreaking to read the transcripts and know that some of the people in the management structure have had to live with the decisions they made.


### Rhetoric as Persuasion

Following Aristotle and more contemporary theorists such as Wayne Booth, I focus on rhetoric as a form of persuasion. Booth's (1974, xiii) definition strikes me as powerful and useful: "Rhetoric is the art of discovering warrantable beliefs and improving those beliefs in shared discourse . . . the art of probing what we believe we *ought* to believe, rather than proving what is true according to abstract methods."

Given the focus on "shared discourse" and "beliefs," I want students to understand how documents fit into the larger social, organizational contexts. One goal is to show them that arguments can be made using many "discursive and textual" tools, and rhetorical analysis "involves a study of the ways in which we attempt to persuade or influence" others (Edwards and Nicoll 2001, 105).

Another important reason for bringing Ray's memorandum as an early example for rhetorical analysis is to point out there is possibly another contributing factor to the *Challenger* disaster that is linked to the failure to communicate but not directly discussed in the literature about the accident thus far: the way the information was presented in print, the document's design. Although much analysis for the *Challenger* disaster focused on "failures in communication that resulted in a decision to launch 51-L based on incomplete and somehow misleading information, a conflict between engineering data and management judgment, and a NASA management structure that permitted internal flight safety problems to bypass Shuttle management" (Presidential Commission 1986, 92), little has been said about the memos themselves as instruments of communication.

Thus, I want students to see that when writing and reading, we must always recognize the texts we create, in print or online, are visual artifacts. In an extremely interesting book about the importance of visual imagery to the future of industry, medicine, education, and virtually every facet of life, Duncan Davies, Diana Bathurst, and Robin Bathurst talk about how and why people communicate (Davies, Bathurst, and Bathurst 1990). They indicate through the use of a picture of a camouflaged moth on the bark of a tree that sometimes communication is interrupted: "In the natural world, some of the most successful species have depended on the interruption of communication or its falsification (to deceive others into acting in the interest of the falsifier), like the plants that persuade insects to carry their genetic material to the right place" (2). Unfortunately, Mr. Ray's memo was like the camouflaged moth; the significant information is hidden in the middle of the memo. As a result, nothing occurred. No actions were taken; nobody was notified of the "results of [his] visits." The information that needed to be received and acted upon was buried in the text; seven people were buried in the ground, effectively grounding the NASA space shuttle program nearly seven years later.

My reason for using Ray's memo is not to condemn anyone. I want to demonstrate that "designers should be concerned about the intended purpose of the artifacts they create and their ultimate effects on society for good or ill" (Bannon 1986, 27). I use the word *designers* to label anyone who is making an attempt to use a structure and adapt it to serve a purpose. Writers of documents are designers, but many are unaware of the visual aspects of the page. They believe writing correct prose is sufficient.

Early in the *Rhetoric*, Aristotle says, "In making a speech one must study three parts: first the means of producing persuasion; second, the style, or language, to be used; third, the proper arrangement of the various parts" (1984, 3.1). A document is rhetorical (Sharf 1979). It must be studied in as much detail as the speeches to which Aristotle refers. One must consider all elements: the method, the language, and the arrangement. Applying visual design has application to both the style and the arrangement. It also, as I implied earlier, has relevance to delivery. Written documents "should be easy to read and therefore easy to deliver [understand]" (Aristotle 1984, 5.3).

Donis Dondis (1973) expresses the significance of this idea when she says, "To expand our ability to see means to expand our ability to understand a visual message and, even more crucial, to make a visual message" (7). This making requires forethought and planning. Audience and aim must be taken into account, and knowledge of how visual elements can contribute to the clarity of the delivery of information is important.

## RHETORICAL ANALYSIS—A FOUNDATION FOR ALL ASSIGNMENTS

Borrowing from both Booth's (1963) concept of the rhetorical stance and, Bitzer's (1968) focus on the rhetorical situation and Brockriede's (1974) concept of rhetorical analysis as argument, I emphasize that rhetorical analysis is an integral step in nearly every assignment or writing task students face in and out of class. Although it may be losing some of its charm, the job-application assignment remains one that students understand (and often appreciate). Many need or soon will need to submit resumes and/or letters to potential employers for internships or jobs.

A good opening strategy is to focus on audience. Aristotle explained that "of the three elements in speech-making—speaker, subject, and person addressed—it is the last one, the hearer, that determines the speech's end and object" (1984, 1.3). We must understand the concept of audience, which, as James E. Porter explains, "is among the slipperiest of rhetoric notions" (Porter 1992, 8). Porter argues that Aristotle's notion of audience has "burden[ed] the rhetor" (15). He spends the rest of his book trying to unburden us by focusing on notions of discourse communities and positioning the audience as "collaborative writer, as a force that shapes and influences the writer" (114). Given the emphasis in our field on such issues as readability and usability, we can help students gain a better perspective of the collaborative nature of writing in the disciplines without having to take them down a lengthy, theoretical road. Perhaps the most important point is "the relation between the organizational situation and the rhetorical situation, and the culture, values, and ways of thinking that determine the criteria for judging communication practice in a real organization" (Driskill 1989, 138).

### Audience Needs and Understanding Genres

In this assignment sequence, a discussion of the audience and an understanding of what that means are critical. You can begin with a specific focus on the practical, highlighting what the audience *needs*, as these needs are outlined in the various job advertisements a company or organization publishes. Thus, the first step is to share recent job ads with students and work collaboratively to ascertain the essential information that will inform their writing. For the purposes of this chapter, I've included two ads, both taken from a search on Idealist.org, and both are within the realm of possibility for a student soon to graduate.

The first step involves examining the genre of job ads, a genre that is not fixed and, as a result, varies widely. Students search for a job in proposal writing[7] on Idealist.org, and we scan and evaluate about ten job ads. Despite different searches due to different semesters, inevitably the students note that the majority of the ads use three main headings: "Job Description," "Responsibilities," and "Qualifications," which is often divided into "minimum" and "preferred." This task helps them understand some issues associated with genre. Students see the expectations, constraints, and flexibility of the genre.

We create a table, in class (or online), based upon the information from the ads. For the two jobs I've included, the table might look like the following:

Table 2.1. Reviewing and assessing job skills

| Job Title | Associate Proposal Writer | Technical Writer |
|---|---|---|
| Minimum qualifications | • BA in English, journalism, communications (or related field)<br>• 1 to 2 years' experience in communications role | • Expertise in English language and grammar<br>• Expertise in MICROSOFT WORD<br>• Ability to explain technical ideas in simple language<br>• Creativity<br>• Ability to self-motive<br>• Ability to work with a team<br>• Competent technical understanding<br>• Ability to ensure technical verbiage is easy to understand by the layperson<br>• Ability to write clear and concise policies and procedures |
| Preferred qualifications | • Excellent writing, editing, communication, and project-management skills with a customer-service orientation<br>• Ability to meet tight deadlines and balance multiple projects<br>• Proficiency with Word and some experience using Excel and PowerPoint<br>• Experience with Qvidian desirable | • Excellent research and writing skills<br>• Understanding of information design, information architecture, training-material development, user interfaces, and business analysis. |
| | • Manage the proposal process, from initial strategy meeting through production, to create a quality product; ensure internal and external clients adhere to Prime's proposal process<br>• Research responses to proposal questionnaires using the Qvidian database, subject-matter experts (SMEs), and other sources; manage the SMEs to ensure they are appropriately informed of the proposal, their role in the process and their timely and complete response to identified questions<br>• Review proposals to identify new questions, areas of concern, unique requirements, and opportunities for differentiation | • Work with engineering and dig into our iOS, Android, and web-based Backend Management Tool Suite to create documentation and test plans<br>• Create and write various types of user documentation, including how-to guides, FAQs, references, manuals, cheat sheets, instructions, and online help; these documents will target technical, business, and consumer audiences<br>• Create operating and maintenance manuals<br>• Assist the sales team in responding to requests for proposals (RFPs), writing white papers, and marketing newsletters |

*continued on next page*

Table 2.1—*continued*

| Job Title | Associate Proposal Writer | Technical Writer |
|---|---|---|
| Key responsibilities or duties | • In partnership with sales, account management, and/or client proposal contacts, develop concise, accurate and well-written responses to proposal questions using the established criteria for quality proposal responses; ensure proposals and supporting materials include strategic positioning and messages as directed by the sales lead, account manager, and/or client<br>• Identify and organize appropriate exhibits to effectively support Prime's proposals; ensure proposal is packaged professionally and meets all proposal requirements and timeframes<br>• Route new and updated responses to proposal content specialist for inclusion in Qvidian<br>• Perform other duties as assigned | |
| Company description | • Prime Therapeutics' fast-paced and dynamic work environment is ideal for proactively addressing the constant changes in today's healthcare industry. Our employees are involved, empowered, and rewarded for their achievements. We value new ideas and work collaboratively to provide the highest quality of care and service to our members.<br>• If you are looking to advance your career within a growing, team-oriented, award-winning company, apply to Prime Therapeutics today and start making a difference in people's lives.<br>• We are proud to be an equal opportunity/affirmative action employer M/F/D/V. We maintain a drug-free and tobacco-free workplace and perform preemployment substance-abuse testing. | • CooCoo launched four years ago and is changing the way people use public transit, allowing riders to pay for their trip on a bus or train using their mobile devices.<br>• CooCoo's platform allows transit riders to plan trips, find out where their buses or trains are in real time, and pay within seconds using their smartphones.<br>• CooCoo has partnered with an industry leader in fare-collection hardware and is currently working on large-scale fare-collection implementations.<br>• CooCoo offers a casual, collaborative work environment, with ample career development and advancement potential.<br>• While still a small team, CooCoo is aggressively growing and is seeking talented, driven, and reliable candidates to round out our technical staff. |
| Company location | • Bloomington, IN<br>• http://www.primetherapeutics.com/index.html | • New York City<br>• https://www.coocoo.com/ |

With this practice analyzing audiences, students gain an additional understanding of style and tone, as well as more exposure to the concept of genre expectations.

*Audience Analysis*

Read any textbook's chapter about job applications, and one of the first and most important "lessons" will be the need to shift from an "I" to a "you" focus. Simply put, students must shift from what a job might mean for them to how their experiences and education can be of value to the employer. While a job ad, in and of itself, cannot give a potential employee a 360-degree picture of a company or its environment, it can give some useful information about *how* the company sees itself or how the company hopes to be seen. Using the two ads, students can glean some interesting perspectives.

Starting with the company descriptions, students can see differences in the companies and how they represent themselves. Prime Therapeutics, a healthcare company, highlights (1) that the work is "fast-paced and dynamic," (2) "team-oriented," and (3) willing to reward employees who are "involved" for their "achievements." They are "award winning," which demonstrates a competitive nature, and while "team-oriented," there is oversight regarding drug and tobacco use, as well as a dedicated (or at least announced) focus on equal opportunity.

CooCoo's description is fascinating for many reasons. Because it is still a new company, the ad provides some history ("launched four years ago"). However, CooCoo's focus is on impact, how it is "changing the way people use public transit." This focus highlights a consumer or civil-society emphasis. The second and third bullets also emphasize looking outward. Not until the fourth bullet does the company describe itself as offering "a casual, collaborative work environment." It is a "small team."

Looking at these clearly articulated self-descriptions can be very helpful toward understanding the kinds of values embedded in the organizational culture. These values also are evident simply by reading the job descriptions. Prime Therapeutics (PT) chooses the header of "Responsibilities," and of the seven responsibilities, six begin immediately with verbs; the seventh has an introductory phrase. The responsibilities ask employees to "manage, research, review, develop, identify, and route." These tasks offer a wide range of opportunities, some of which are creative in nature (although the most creative—"develop"—is done "in partnership with" others).

CooCoo's ad has a very different feel from the heading on. Beginning with the heading of "Description," the ad includes responsibilities, but they are framed in a narrative. The narrative is not visible in the table, but some of the differences are quite visible, specifically the emphasis on creativity. The "Description" uses a form of the verb *to create* three times.

Combining this emphasis on the ability "to create" with the "core skill" of being "creative" makes their culture and the position's framing very different than the one at PT.

### The Roles of Style and Tone

By looking closely at differences between the two companies' perspectives on employee responsibilities, we've already begun an analysis of style and tone. The narrative focus of CooCoo's ad has already been noted, as well as the differences in verbs. But the tone, too, is evident in the overall structure. PT follows a much more traditional model. Perhaps the difference is due to PT's being a more established company with a more mature organizational culture; perhaps it is due to the field—healthcare; perhaps it is the location—Bloomington, Indiana. Regardless, the style, which in this case includes the organizational structure, points to a difference worth noting.

The noted rhetorician Carolyn Miller quotes Aristotle's focus on rhetoric as dealing with "things about which we deliberate but for which we have no systematic rules" and argues that in "more contemporary language rhetoric concerns uncertainty" (Miller 1989, 43). Job ads are among the "things" students deal with. Some will have had successful internships or other activities that have launched them on a trajectory for employment. But many will not have that foundation, and the prospect of what they will do when they graduate can appear daunting. This exercise provides analytical skills and practice coping with uncertainty. Equally important, because job ads are versatile, students can find ads in their own fields that can help them see commonality and divergence, which opens up opportunities to discuss the social nature of language, particularly specialized disciplinary language.

## GENRE ANALYSIS AND DISCIPLINARY KNOWLEDGE

Up to now, my goal has been two-fold: (1) to provide an argument outlining the foundational importance of rhetorical analysis as an integral part of every course and (2) to demonstrate its value using two genres that factor into most professional and technical writing courses: memos (correspondence) and the texts involved in moving from the world of higher education to the world of work (job ads). As students work their way through these assignments, you can also help them learn critical information about relevant questions: What counts as evidence? What kinds of assumptions are shared within discourse communities (such as the one at NASA involving the engineers)? How, if at all, do writers in various disciplines use or modify standard genres such as the memo? What differences in tone and style exist among disciplines? As they consider these questions, they engage in activities that have roots in classical rhetorical education in which advanced study required the study and imitation of various models.

These questions will help you initiate larger discussions about genres. One way is to ask students to find examples of memos linked to workplaces in their disciplines and share them with students from other disciplines. Doing so may be useful in helping students understand that "genres are groups of discourses which share substantive, stylistic, and situational characteristics" (Campbell and Jamieson 1978, 20). By focusing on documents in students' home disciplines, you engage them in discussions that further drive home the point that "genres have been defined by similarities in strategies or forms in the discourse, by similarities in audience, by similarities in modes of thinking, by similarities in rhetorical situations" (Miller 1984, 151). Students begin to see that recognizing these similarities will help them become professionals, as they will begin to recognize structures and strategic choices writers in their chosen fields make. Gerard Hauser (2002) explains how these "structures . . . function strategically to shape our attitudes, beliefs, and actions" (249). Finally, asking students to focus on whether or not the documents they share are persuasive or convincing can lead to discussions about questions mentioned above, such as, What counts as evidence in specific fields?

By combining a discussion about what they feel makes the document convincing (or not convincing) and framing it in the larger context of the document's relevance, students can see the importance and places of writing and rhetoric in their fields. This follow-on assignment also helps students understand that in virtually every discipline, routine documents can have significant or potentially significant consequences. One need only look at the memos and e-mail from Microsoft's antitrust lawsuit or Hillary Clinton's staff to drive home this point.

## CONCLUSION

Early in my career, I wrote about the process I went through to build a program in professional writing at Virginia Tech. In the earliest published piece about that process, based upon a presentation at the Council for Programs in Technical and Scientific Communication (1999), I argued that the capstone course should be one focusing on rhetorical analysis, and I outlined some initial course objectives:

- To provide students with a better understanding of the roles language plays in the creation of knowledge;
- To help students conceive of the "culture" that derives from their "special work" (Earle, quoted in Kynell 33);
- To help students in this field learn to function more fully as citizens. (23)

All of these objectives are derived from a foundational belief in the link between rhetoric and culture, expressed beautifully by Karlyn K. Campbell (2006) when she argues that "rhetoric is ubiquitous. Never ask if there is rhetoric; where there is culture and language, there is rhetoric. The challenge is to discover its cultural forms and functions" (358). These objectives also were built upon a belief that teaching rhetorical analysis is truly a "pedagogical imperative" (Medhurst 1989, 176), part of what brings our texts to life and enables us to help our students shift from ignorance to knowledge through reasoning as a result of rhetorical analysis, to achieve the "disposition" Dunne (1993) describes. Such a disposition is the result of guided practical experience, a formal training that will lead to knowing how to learn rather than learning how to write using formulaic means.

In this chapter, I've offered strategies teachers can use in virtually any foundational course in professional and technical communication. These rely on underlying questions regarding the disposition of a rhetorical scholar. While there are many sets of questions I could draw upon, all valuable and many equally as good, I like the ones Michael Medhurst (2013) outlines in his recent discussion about rhetorical critics.

> The first task of rhetorical critics is to properly understand the nature of the art form before them. What kind of thing is it? (a question of genre). For what purpose was the thing created? (a question of motive). For whom is the message created? (a question of audience). How is this thing like or unlike others of its type? (a question of typicality or uniqueness). By what strategy or technique does the art form operate? (a question of internal movement). How are those strategies or techniques related to the audience that seems to be intended? (a question of audience adaptation). How well did the creator of the thing utilize the materials at his or her disposal in relation to the audience and intended purpose? (a question of rhetorical execution). (126–27)

Using these questions when confronting any text, whether it is visual or verbal, online or on air, in a website or long treatise, will provide the means to help that text "speak" about the person or entity that created it, the purpose behind its creation, and some information about the role it plays in a larger social context Sharf 1979 and Waddell 2000. We can't guarantee that using these questions (or whatever set a teacher prefers) as a heuristic will make all students rhetors in the sense Quintilian sought to do, but we can help them learn more about strategies people employ when using symbols to communicate.

## DISCUSSION QUESTIONS

1. In this chapter, we examined two very different company websites to help us more fully understand their organizational cultures. Knowing more about organizational cultures often can be of use when you apply for a position. Another strategy is to look at the company's press releases and annual reports. What do you learn when you review these documents on the Prime Therapeudics and CooCoo websites? Does it confirm or expand what you learned from analyzing the job ads?

2. Examine websites of two competing international transportation companies (UPS and DHL) or, in a very different part of the corporate world, Winbro ( www.winbrogroup.com) and Makino ( www.makino. com). What similarities do you find in terms of the language they use to define themselves? How do they try to set themselves apart? What do these insights tell you about the companies and how they "see" themselves and their audiences?

3. Choose ten Facebook profiles at random; what do the profiles say about the writers and the writers' sense of audience? Focus on two sections: "About" and "Photos." What kinds of "claims" do the sections make about the individuals? What can we learn about them by thinking about design and tone?

4. One of the more controversial topics in the past several decades has been the concept of "cold fusion." It is purportedly a means of generating nuclear energy without dangerous radiation. Examine this issue by focusing on several sources: http://coldfusionnow.org/ and https://ae on.co/essays/why-do-scientists-dismiss-the-possibility-of-cold-fusion. What can you tell about the authors' perspectives of their audiences?

# APPENDIX 2.A. NASA MEMO

National Aeronautics and
Space Administration

George C. Marshall Space Flight Center
Marshall Space Flight Center, Alabama
35812

*BEN*

**NASA**

EP25 (79-23)                                      February 6, 1979

TO:          Distribution

FROM:        EP25/Mr. Ray

SUBJECT:     Visit to Precision Rubber Products Corporation and
             Parker Seal Company

The purpose of this memorandum is to document the results of a visit
to Precision Rubber Products Corporation, Lebanon, TN, by Mr. Eudy, EE51 and
Mr. Ray, EP25, on February 1, 1979 and also to inform you of the visit
made to Parker Seal Company, Lexington, KY on February 2, 1979 by Mr. Ray.
The purpose of the visits was to present the O-ring seal manufacturers
with data concerning the large O-ring extrusion gaps being experienced on
the Space Shuttle Solid Rocket Motor clevis joints and to seek opinions
regarding potential risks involved.

The visit on February 1, 1979, to Precision Rubber Products Corporation
by Mr. Eudy and Mr. Ray was very well received. Company officials, Mr.
Howard Gillette, Vice President for Technical Direction, Mr. John Hoover,
Vice President for Engineering,and Mr. Gene Hale, Design Engineer
attended the meeting and were presented with the SRM clevis joint seal
test data by Mr. Eudy and Mr. Ray. After considerable discussion,
company representatives declined to make immediate recommendations because
of the need for more time to study the data. They did; however, voice
concern for the design,stating that the SRM O-ring extrusion gap was
larger than that covered by their experience. They also stated that more
tests should be performed with the present design. Mr. Hoover promised
to contact MSFC for further discussions within a few days. Mr. Gillette
provided Mr. Eudy and Mr. Ray with the names of two consultants who may
be able to help. We are indebted to the Precision Rubber Products
Corporation for the time and effort being expended by their people in
support of this problem, especially since they have no connection with
the project.

The visit to the Parker Seal Company on February 2, 1979, by Mr. Ray,
EP25, was also well received; Parker Seal Company supplies the O-rings
used in the SRM clevis joint design. Parker representatives, Mr. Bill
Collins, Vice President for Sales, Mr. W. B. Green, Manager for Technical
Services, Mr. J. W. Kosty, Chief Development Engineer for R&D, Mr.
D. P. Thalman, Territory Manager and Mr. Dutch Haddock, Technical
Services, met with Mr. Ray, EP25, and were provided with the identical

SRM clevis joint data as was presented to the Precision Rubber Products
Company on February 1, 1979. Reaction to the data by Parker officials
was essentially the same as that by Precision; the SRM O-ring extrusion
gap is larger than they have previously experienced. They also expressed
surprise that the seal had performed so well in the present application.
Parker experts would make no official statements concerning reliability
and potential risk factors associated with the present design; however,
their first thought was that the O-ring was being asked to perform beyond
its intended design and that a different type of seal should be considered.
The need for additional testing of the present design was also discussed
and it was agreed that tests which more closely simulate actual conditions
should be done. Parker officials will study the data in more detail with
other Company experts and contact MSFC for further discussions in
approximately one week. Parker Seal has shown a serious interest in
assisting MSFC with this problem and their efforts are very much appreciated.

*William L. Ray*
William L. Ray
Solid Motor Branch, EP25

Distribution:
SA41/Messrs. Hardy/Rice
EE51/Mr. Eudy
EP01/Mr. McCool

# APPENDIX 2.B. JOB AD FOR PRIME THERAPEUTICS

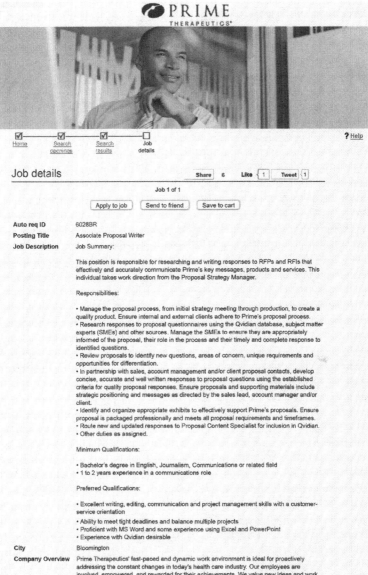

Home — Search openings — Search results — Job details

? Help

## Job details

Share 6    Like 1    Tweet 1

Job 1 of 1

[ Apply to job ]    [ Send to friend ]    [ Save to cart ]

| | |
|---|---|
| **Auto req ID** | 6028BR |
| **Posting Title** | Associate Proposal Writer |
| **Job Description** | Job Summary: |

This position is responsible for researching and writing responses to RFPs and RFIs that effectively and accurately communicate Prime's key messages, products and services. This individual takes work direction from the Proposal Strategy Manager.

Responsibilities:

• Manage the proposal process, from initial strategy meeting through production, to create a quality product. Ensure internal and external clients adhere to Prime's proposal process.
• Research responses to proposal questionnaires using the Qvidian database, subject matter experts (SMEs) and other sources. Manage the SMEs to ensure they are appropriately informed of the proposal, their role in the process and their timely and complete response to identified questions.
• Review proposals to identify new questions, areas of concern, unique requirements and opportunities for differentiation.
• In partnership with sales, account management and/or client proposal contacts, develop concise, accurate and well written responses to proposal questions using the established criteria for quality proposal responses. Ensure proposals and supporting materials include strategic positioning and messages as directed by the sales lead, account manager and/or client.
• Identify and organize appropriate exhibits to effectively support Prime's proposals. Ensure proposal is packaged professionally and meets all proposal requirements and timeframes.
• Route new and updated responses to Proposal Content Specialist for inclusion in Qvidian.
• Other duties as assigned.

Minimum Qualifications:

• Bachelor's degree in English, Journalism, Communications or related field
• 1 to 2 years experience in a communications role

Preferred Qualifications:

• Excellent writing, editing, communication and project management skills with a customer-service orientation
• Ability to meet tight deadlines and balance multiple projects
• Proficient with MS Word and some experience using Excel and PowerPoint
• Experience with Qvidian desirable

| | |
|---|---|
| **City** | Bloomington |
| **Company Overview** | Prime Therapeutics' fast-paced and dynamic work environment is ideal for proactively addressing the constant changes in today's health care industry. Our employees are involved, empowered, and rewarded for their achievements. We value new ideas and work collaboratively to provide the highest quality of care and service to our members. |

If you are looking to advance your career within a growing, team-oriented, award-winning company, apply to Prime Therapeutics today and start making a difference in people's lives.

We are proud to be an Equal Opportunity/Affirmative Action Employer M/F/D/V. We maintain a drug-free and tobacco-free workplace and perform pre-employment substance abuse testing.

No Agencies.

# APPENDIX 2.C. JOB AD FOR COOCOO

**indeed**
one search. all jobs.

what: [                    ]
job title, keywords or company

where: [ New York, NY ]    [ Find Jobs ]    Advanced Job Search
city, state, or zip

## Technical Writer
CooCoo - New York, NY

» **Apply Now**
Please review all application instructions before applying.

**Technical Writer, Product**

[ Follow ]    Get job updates from CooCoo

**Company**

CooCoo launched four years ago and is changing the way people use public transit - allowing riders to pay for their trip on a bus or train using their mobile device. CooCoo's platform allows transit riders to plan trips, find out where their bus or train is in real-time and pay within seconds using their smart phone. CooCoo has partnered with an industry leader in fare collection hardware and is currently working on large scale fare collection implementations.

CooCoo offers a casual, collaborative work environment, with ample career development and advancement potential. While still a small team, CooCoo is aggressively growing and is seeking talented, driven, and reliable candidates to round out our technical staff.

**Description**

You will work with our engineering and dig into our our iOS, Android and Web-based Backend Management Tool Suite to create documentation and test plans. This position will be responsible for creating and writing various types of user documentation, including how-to guides, FAQs, references, manuals, cheat sheets, instructions and online help.

These documents will target technical, business and consumer audiences. Operating and maintenance manuals are also created by the technical writer. These are created for the technical worker who has a greater understanding of the subject.

This candidate will also assist the sales team in responding to Request for Proposals (RFPs), writing white papers and marketing newsletters.

To fill this position, you should have excellent research and writing skills, along with an understanding of information design, information architecture, training material development, user interfaces, and business analysis.

**Core Skills**

- Expert in English Language and grammar
- Microsoft Word
- Explain technical ideas in simple language
- Creative
- Self motivator
- Working with a team
- Competent technical understanding
- Ensure technical verbiage is easy to understand by the layperson
- Write clear and concise policies and procedures

**Bonus Skills**

- Adobe Creative Suite

Indeed - 3 days ago - save job

» **Apply Now**
Please review all application instructions before applying.

[ Apply Now ]

## Notes

1.   See Vocabulary.com: "*Rhetoric* is the art of written or spoken communication. If you went to school a hundred years ago, your English class would have been called Rhetoric. But nowadays if we say something is rhetorical, we usually mean that it's only good for talking."
2.   1967 fire of Apollo 1 with Virgil I. "Gus" Grissom, Roger Chafee, and Edward White II.
3.   Howard Gardner is an American developmental psychologist from Harvard University whose theories of multiple intelligences inform a range of disciplines.
4.   Whether they highlight the importance of understanding the reader's needs and values (Anderson 2014, X) or whether, by presenting a range of sample documents, they emphasize the rhetorical situation as a whole (Markel 2014, 10), nearly every professional writing textbook opens with a focus on the rhetorical process.
5.   See http://rhetoric.byu.edu/Pedagogy/Rhetorical%20Analysis%20heuristic.htm or http://www.wikihow.com/Write-a-Rhetorical-Analysis
6.   See http://mhhe.com/mayfieldpub/tsw/rep-trip.htm.
7.   Any relevant job will work for this part of the exercise.

## References

Anderson, Paul. 2014. *Technical Communication: A Reader-Centered Approach*. 8th ed. Boston, MA: Cengage Learning.

Aristotle. 1984. *The Rhetoric and Poetics of Aristotle*. New York: Modern Library.

Arnold, Vanessa D., and John C. Malley. 1988. "Communication: The Missing Link in the Challenger Disaster." *Business Communication Quarterly* 51 (1): 12–14.

Bannon, Liam J. 1986. "Issues in Design: Some Notes." In *User Centered System Design*, edited by Donald Norman and Stephen W. Draper, 25–29. Hillsdale, NJ: Lawrence Erlbaum.

Bernhardt, Stephen. 1986. "Seeing the Text." *College Composition and Communication* 37 (1): 66–78.

Bitzer, Lloyd. 1968. "The Rhetorical Situation." *Philosophy and Rhetoric* 1 (1): 1–14.

Boisjoly, Russell P., Ellen F. Curtis, and Eugene Mellican. 1989. "Roger Boisjoly and the Challenger Disaster: The Ethical Dimensions." *Journal of Business Ethics* 8 (2): 217–30.

Booth, Wayne. 1963. "The Rhetorical Stance." *College Composition and Communication* 14 (3): 139–45.

Booth, Wayne. 1974. *Modern Dogma and the Rhetoric of Assent*. Notre Dame, IN: University of Notre Dame Press.

Booth, Wayne. 2004. *The Rhetoric of Rhetoric*. Malden, MA: Blackwell.

Brassring.com. 2015. Job ad for Technical Writer at Indeed.com. Brassring.com. Accessed Feb 1, 2015.

Brockriede, Wayne. 1974. "Rhetorical Criticism as Argument." *Quarterly Journal of Speech* 60 (2): 165–74.

Campbell, Karlyn K. 2006. "Cultural Challenges to Rhetorical Criticism." *Rhetoric Review* 25 (4): 358–61.

Campbell, Karlyn K., and Kathleen H. Jamieson. 1978. "Form and Genre in Rhetorical Criticism: An Introduction." In *Form and Genre: Shaping Rhetorical Action*, edited by Karlyn K. Campbell and Kathleen H. Jamieson, 9–32. Falls Church, VA: Speech Communication Association.

Ceccarelli, Leah. 2001. *Shaping Science with Rhetoric*. Chicago, IL: University of Chicago Press.

Corbett, Edward P. J. 1971. "The Theory and Practice of Imitation in Classical Rhetoric." *College Composition and Communication* 22 (3): 243–50.

Covault, Craig. 1986. "O-Ring Documentation Misled Managers." *Aviation Week and Space Technology* 124 (12): 26.

Crane, Milton. 1956. "An Exercise in the Teaching of Rhetorical Analysis." *Journal of General Education* 8 (4): 226–30.

Cyphert, Dale. 2010. "The Rhetorical Analysis of Business Speech." *Journal of Business Communication* 47 (3): 346–68.

Davies, Duncan, Diana Bathurst, and Robin Bathurst. 1990. *The Telling Image*. Oxford: Clarendon.

Dombrowski, Paul M. 1992. "Challenger and the Social Contingency of Meaning: Two Lessons for the Technical Communication Classroom." *Technical Communication Quarterly* 1 (3): 73–86.

Dondis, Donis A. 1973. *A Primer of Visual Literacy*. Cambridge: MIT Press.

Driskill, Linda. 1989. "Understanding the Writing Context in Organizations." In *Writing in the Business Professions*, edited by Myra Kogan, 125–45. Urbana, IL: NCTE and ABC.

Dubinsky, James. 1993. "The Rhetoric of Page Design." Paper presented at the National Council of Teachers of English Conference, Pittsburgh, PA, November 1993. Eric Document No. 371211.

Dubinsky, James. 1999. "Mending Walls and Adding Gates: Intradisciplinary Collaboration and Building Programs that Work." In *Proceedings of the 26th Council for Technical and Scientific Communication*, edited by Carolyn Rude, 20–28. Santa Fe, NM: Council for Programs in Technical and Scientific Communication.

Dubinsky, James. 2004. "Becoming a User-Centered Reflective Practitioner." *Teaching Technical Communication: Critical Issues for the Classroom*, edited by James Dubinsky, 1–8. Boston, MA: Bedford/St. Martin's.

Dunne, Joseph. 1993. *Back to the Rough Ground: Phronesis and Techne in Modern Philosophy and in Aristotle*. Notre Dame, IN: University of Notre Dame Press.

Edwards, Richard, and Katherine Nicoll. 2001. "Researching the Rhetoric of Lifelong Learning." *Journal of Education Policy* 16 (2): 103–12.

Genung, John F. 1888. *Handbook of Rhetorical Analysis*. Boston, MA: Ginn.

Hauser, Gerard. 2002. *Introduction to Rhetorical Theory*. 2nd ed. Long Grove, IL: Waveland.

Herndel, Carl, Caroline Miller, and Bruce Fennell. 1991. "Understanding Failures in Organizational Discourse: The Accident at Three Mile Island and the Space Shuttle Challenger Disaster." In *Textual Dynamics of the Professions: Historical and Contemporary Writing in Professional Communities*, edited by Charles Bazerman and John Paradis, 279–305. Madison: University of Wisconsin Press.

Huckin, Thomas N. 1997. "Critical Discourse Analysis." In *Functional Approaches to Written Text: Classroom Applications*, edited by Thomas Miler, 78–93. Washington, DC: United States Information Agency.

Indeed.com. 2015. Technical Writer at CooCoo. Indeed.com. Accessed Feb 1, 2015.

Joseph, Sister Miriam. 2002. *The Trivium: The Liberal Arts of Logic, Grammar, and Rhetoric*. Philadelphia, PA: Paul Dry Books.

Luke, Allan. 2002. "Beyond Science and Ideology Critique: Developments in Critical Discourse Analysis." *Annual Review of Applied Linguistics* 22: 96–110.

Markel, Michael. 2014. *Technical Communication*. 11th ed. Boston, MA: Bedford St. Martin's.

McIlavine, J. 1874. "Rhetorical Analysis and Synthesis." *Presbyterian Quarterly and Princeton Review* 3 (11): 456–83.

McKerrow, Richard. 2013. "The Critical Impulse." In *Purpose, Practice, and Pedagogy*, edited by James A. Kuypers, 109–22. Lanham, MD: Lexington Books.

Medhurst, Michael. 1989. "Teaching Rhetorical Criticism to Undergraduates: Special Editor's Introduction." *Communication Education* 38 (3): 175–77.

Medhurst, Michael. 1998. "Rhetorical Education in the Twenty-First Century." *Southern Communication Journal* 63 (4): 346–49.

Medhurst, Michael. 2013. "Rhetorical Criticism as Textual Interpretation." In *Purpose, Practice, and Pedagogy*, James A. Kuypers, 123–35. Lanham, MD: Lexington Books.

Mehlenbacher, Brad. 2013. "What Is the Future of Technical Communication?" In *Solving Problems in Technical Communication*, edited by Johndan Johnson-Eilola and Stuart A. Selber, 187–208. Chicago, IL: University of Chicago Press.

Miller, Caroline. 1984. "Genre as Social Action." *Quarterly Journal of Speech* 70 (2): 151–67.

Miller, Caroline. 1989. "The Rhetoric of Decision Science, or Herbert A. Simon Says." *Science, Technology, and Human Values* 14 (1): 43–46.

Ong, Walter. 1971. *Rhetoric, Romance, and Technology*. Ithaca, NY: Cornell University Press.

Perelman, Les, James Paradis, and Edward Barrett. 2001. *The Mayfield Handbook for Technical and Scientific Writing*.

Porter, James E. 1992. *Audience and Rhetoric*. Englewood Cliffs, NJ: Prentice Hall.

Presidential Commission on the Space Shuttle Accident. 1986. *Report to the President*. 5 vols. Washington, DC: Government Printing Office.

Romzek, Barbara S., and Melvin J. Dubnick. 1987. "Accountability in the Public Sector: Lessons from the Challenger Tragedy." *Public Administration Review* 47 (3): 227–36.

Rowland, Robert A. 1986. "The Relationship between the Public and the Technical Spheres of the Argument: A Case Study of the Challenger Seven Disaster." *Central States Speech Journal* 37 (3): 136–46.

Russell, David R. 1997. "Rethinking Genre in School and Society: An Activity Theory Analysis." *Written Communication* 14 (4): 504–54.

Schriver, Karen. 2013. "What Do Technical Communicators Need to Know about Information Design?" In *Solving Problems in Technical Communication*, edited by Johndan Johnson-Eilola and Stuart A. Selber, 386–427. Chicago, IL: University of Chicago Press.

Sharf, B. F. 1979. "Rhetorical Analysis of Nonpublic Discourse." *Communication Quarterly* 27 (3): 21–30.

Spinuzzi, Clay. 2004. *Tracing Genres through Organizations: A Sociocultural Approach to Information Design*. Cambridge, MA: MIT Press.

Spinuzzi, Clay. 2013. "How Can Technical Communication Study Work Contexts?" In *Solving Problems in Technical Communication*, edited by Johndan Johnson-Eilola and Stuart A. Selber, 262–84. Chicago, IL: University of Chicago Press.

Turner, Kathleen J. 2013. "The Glory of Rhetorical Analysis." In *Purpose, Practice, and Pedagogy*, edited by James A. Kuypers, 177–91. Lanham, MD: Lexington Books.

van Maanen, John. 2011. *Tales of the Field: On Writing Ethnography*, 2nd ed. Chicago: University of Chicago Press.

van Manen, Max. 1994. "Pedagogy, Virtue, and Narrative Identity." *Curriculum Inquiry* 24 (2): 135–70.

Vocabulary.com. "Rhetorical." Accessed November 11, 2016.

Waddell, Craig. 2000. *And No Birds Sing: Rhetorical Analyses of Rachel Carson's Silent Spring*. Carbondale: Southern Illinois University Press.

Winsor, Dorothy A. 1988. "Communication Failures Contributing to the Challenger Accident." *IEEE Transactions on Professional Communication* 3 (3): 101–7.

Yates, Joanne, and Wanda Orlikowski. 2002. "Genre Systems: Structuring Interaction through Communicative Norms." *Journal of Business Communication* 39 (1): 13–35.

# 3

# TEACHING STUDENTS ABOUT STYLE IN TECHNICAL COMMUNICATION

Dan Jones

You have just been assigned to teach a section of the introductory technical communication course. Initially, you may feel overwhelmed by the challenges—for example, teaching about instructions, technical descriptions, technical reports, proposals, policies and procedures, technical illustrations, and technical presentations. Yet you also know the course focuses on the essentials of good writing, and you take some comfort in knowing you will be teaching about the writing process, audience analysis, the rhetorical situation, and the modes of persuasion—ethos, pathos, and logos. You also note at least part of the course will focus on elements of style, including active and passive voice, tone, clarity, and conciseness. Although you may not know everything you need to know to teach the course, you know enough for a good start, and you know you will learn the rest along the way.

Once you begin teaching the course, you discover many students can write competently but not as clearly, concisely, courteously, or persuasively as they should. Some paragraphs in their documents continue to lack good topic sentences, some paragraphs are much too long, or some are not well developed or complete. Some sentences remain ambiguous, other sentences are too wordy, and the flow from sentence to sentence could be improved in places. In some documents, the tone should still be more professional or courteous, while in other documents, some students may demonstrate good reasoning or provide substantial evidence, but the work remains unconvincing. Or perhaps, more than any of the above, you would just like to see more students write more engaging prose, prose demonstrating their particular personalities, especially in their correspondence. Of course, you expect many of these weaknesses. It's only an introductory course after all. The introductory technical communication textbook you use devotes only a brief chapter to the basics, but you know you must cover issues of style

DOI: 10.7330/9781607326809.c003

more in depth to help students become better writers. You wonder what you can do.

In this chapter, I emphasize that what we teach students about style can make a difference in their success on the job. We must devote more attention to matters of style than most of us likely do or have the time to do. Of course, style can be taught to students in a variety of ways, and most of us only have time to cover the basics. However, I argue here for a more in-depth approach to teaching style, whether in the introductory technical communication course or a more advanced course. The complexity of communication in the workplace requires a comprehensive approach. I begin with an overview of some of the key literature discussing the importance of teaching style in the technical communication classroom and review how teaching students more about style has gained wider acceptance. I then review both basic and more in-depth approaches to teaching style and provide some examples of possible exercises and other assignments.

## TEACHING STYLE THEN AND NOW

The most helpful literature discussing the importance of teaching style in technical communication and strategies for teaching it largely spans the past six decades. Randolph Hudson (1961) reviewed the challenges for teaching technical writing to engineers at a time when many in the humanities were tasked to teach this group with little specialized preparation for doing so. He argues for applying some novel techniques for the time, such as abandoning the traditional research paper "because it offers little practice for any skills which engineers will employ in their professional lives" (210). One of his particular concerns is how the widespread use of technical terminology in the technical professions presents special challenges. Hudson comments, "A teacher cannot in clear conscience ban technical terminology. . . . A teacher must accept the fact that his students are learning to master their technical vocabularies, and he should encourage them to seek out the precise words which have been devised to describe specific referents" (212).

Charles Stratton (1979) argues nonexperts can be taught to discriminate between excellent technical writing and less excellent writing, students can be taught to recognize measurable differences between these two, and students can improve their writing in significant ways by focusing on stylistic characteristics of prose (4).

Others call our attention to how play or experimenting with or even having fun with language in our communications can often best serve the interests of both the writer and the audience. Richard Lanham (1974, 1983, 2000) suggests we expand our notions of the domain of style to include not only purpose but also play and competition. Lanham (1983) believes that "the really persuasive people have an instinctive grasp of purpose, game, and play. And it is precisely this mixture that prose style in its fullness always expresses, and that prose analysis can teach" (9). Lanham suggests a study of prose style helps us know "*how* to use information" and "*what kind of message* a message is" (10). Stephen Bernhardt (1994) acknowledges Lanham's influence and reinforces an emphasis on play: "English classes should attend to style because it is fun. Style is the locus of play in languages. . . . Playing with style allows us to lighten up, to take risks, to experiment with new ways of saying things. Good writing is leavened with the pleasure the writer takes in matching style to the occasion" (181).

Becky Bradway (1997) was puzzled about the neglect of teaching style in writing courses, or at least she was surprised that some think style deserves emphasis in creative writing and literature classes but not as much in technical writing classes. After working as a writer in industry for twelve years, she returned to the classroom in graduate school assuming teaching style would be equally valued by the different groups in her department. She comments, "Style was, to me, one of the most important—the most basic and central—elements of writing. But there was very little or no discussion of style . . . . The subject was dismissed as unimportant, at the least, and intrusive or dangerous or irrelevant at worst" (22). She reminds us, "The development of individual style, suitable to situations of audience, reflecting personal commitments and personality, makes for better writing in any community" (30).

More recently, Brian Blackburne's (2014) approach for teaching style in technical writing courses is one of the most helpful. He uses the specific genre of airline-safety briefings, providing students a framework for analyzing the style of these common documents. Students see how many documents typically fail to meet the principles of effective technical communication, particularly those based on Joseph Williams's (2013) study of style. Blackburne concludes with a helpful list of actions and their rationales we can use in our technical communication courses. He suggests, "Because students will enter a variety of professional situations already entrenched in accepted writing styles, the greatest hope we have for improving the state of workplace writing is to graduate students who have a working knowledge of style and its effects on everyday documents" (83).

As for introductory technical communication textbooks, they continue to improve in their coverage of style. In the first edition of their textbook, Kenneth Houp and Tom Pearsall provide a chapter discussing some of the basics of technical writing style (Houp and Pearsall 1968). The third edition of *Modern Technical Writing* (Sherman and Johnson 1975), the technical communication textbook I used when I taught my first introductory technical communication courses in the spring of 1980, also provides a chapter on effective style, discussing the effect of diction and sentence structure on style as well as on conciseness. However, the discussions of style are relatively brief in these early textbooks. By comparison, most of today's major and most widely used comprehensive textbooks for introductory technical communication courses provide more thorough discussion, such as those by John Lannon and Laura Gurak (Lannon and Gurak 2013), Mike Markel (2012), Paul Anderson (2013) and Steven Gerson and Sharon Gerson (Gerson and Gerson 2013).

It seems we agree more than ever on the benefits of teaching students strategies for improving their prose style in the technical writing classroom.

## TEACHING THE BASICS OF STYLE

Before students can work their way through different style exercises or assignments, they must have a good foundation in the essentials of effective prose style, from discussing definitions of style to discussing the importance of style choices at the word, sentence, and paragraph levels for a particular audience. Most introductory technical communication textbooks—whether they are comprehensive or more concise—take a basic approach focusing chiefly on the essentials. In their more concise introductory textbook, Sam Dragga and Elizabeth Tebeaux discuss the paragraph, basic principles (determining the reader's knowledge of the subject; determining whether a particular style will be expected; adjusting the style to the readers, the purpose, and the context), keys to building effective sentences (watching sentence length; keeping subjects and verbs close together; writing squeaky-clean prose; avoiding pompous language; writing to express, not to impress; avoiding excessive use of *is/ are* verb forms; using active voice for clarity; and word choice) (Dragga and Tebeaux 2015, 57–73). Dragga and Tebeaux provide one of the best overviews of the basics of style, even when compared to the style chapters of more comprehensive textbooks.

In the introductory course, most of us require students to complete exercises on some of the basics of style, then we quickly move on to other modules, such as writing reports or proposals. We supplement the definition of style provided in the textbook with additional definitions we provide or additional definitions we require students to find and then discuss. We assign exercises on words—abstract and concrete, specific and general, denotation and connotation, jargon, commonly confused words, and more. We work with them on abbreviations and acronyms. We provide exercises on sentences, requiring students to make them clearer or more concise. We require students to identify sentences written in the active voice and those written in the passive voice, and we explain that at times using the passive voice is fine. All these basic exercises and others have their value.

We can, however, go beyond these simple exercises and require an in-depth style analysis of a professional document, as Blackburne (2014) does for his classes, or create our own in-depth assignments. For example, we can require students to create a memo full of affected prose for a well-defined audience and then require a thorough revision into much clearer prose. We can provide sample documents with pretentious prose and ask them to revise the documents into a plain style. Or we can require them to create a memo in a plain style and then revise it into a pretentious style. We can ask students to compose memos or e-mails (or find professional examples) with too much abstract language and then require them to revise the work into more concrete language, discussing the revisions they make and why. We can ask them to find or create a short jargon-filled document in their field, have them revise the work into plain language, and then discuss why doing so may or may not work depending on the rhetorical context.

A role-playing assignment might have particular value for reinforcing the basics of style and tone. Think of the following assignment as a variation of the one used in some introductory courses requiring students to write a letter of complaint and then to respond with a letter of adjustment. You may first need to provide an overview of the elements of these kinds of correspondence and review how certain word choices (for example, *your fault, your mistake*) will be perceived by the audience. You may also need to review strategies for establishing a "you attitude," or seeing and discussing ideas from the audience's point of view.

For just one possible approach, divide a class of twenty students into five groups with four students in each group. (Larger groups can work as well.) Ask each group to create two small businesses or companies, providing each company a name and a product line (for example, computers, cell phones, car audio systems) or services (for instance, repair, maintenance, consulting, or supplies). Then ask two students in the group to role play as employees of one of the companies while the other two role play as employees of the other company. Ask them to create their own job titles and perhaps brief job descriptions. Consider asking them to adopt a particular company role as well—for example, the company curmudgeon, the highly experienced senior staff member, the department manager, or the software tool-savvy employee.

Then ask them to brainstorm about possible contentious scenarios— a missed deadline or missed delivery, poor service, or a faulty product are just a few possibilities. Then ask these employees (students) in each group to collaborate on writing two complaints as e-mails or memos or letters to the other company about this service or product issue, but require them to establish entirely different tones and prose styles in the two communications on this same topic. One communication should convey an angry and unprofessional tone and employ an informal prose style while the other should be professional and courteous throughout and use a more formal prose style. After the two groups share their communications with each other, ask members of each group to collaborate in responding with a communication of adjustment to one of the other group's two communications, denying or refusing what is requested from the other and establishing a tone and prose style they think is most appropriate for the occasion. Challenge the students to be imaginative and specific about the products or services, dates, invoices, events, personnel involved, and more. The communications of the two groups should be appropriately detailed.

This role-playing exercise may take the entire class period or even longer. Much depends on the abilities and the imaginations of students, and much depends on helping any group that has questions or concerns. Once the six communications (two complaints by each two-person group and one communication of adjustment by each two-person group) are completed, ask each group to discuss what they have learned first within these small groups. Then hold a larger class discussion of what the students have learned about writing for a particular audience and adopting a particular tone and style for their different purposes. As much as possible, focus the discussion on particular word choices and the effects of these choices, particularly in terms of how they affect the tone and professionalism of each communication.

We can also explore ways to help students experiment more with paragraphs. Many technical communication textbooks pay some attention to style at the paragraph level, discussing and illustrating, for example, the qualities of a good paragraph (a clear topic sentence, unity, completeness, coherence, and order). Of course, all these qualities are basic, but providing various assignments requiring students to recognize these qualities in selected paragraphs and create sample paragraphs with these qualities can be invaluable, both for the writing the students still must complete in the introductory course and for all the writing they will do later in other college courses or in industry.

Exercises on paragraphs are simple. Just find excellent examples and discuss structure with students (or find examples in need of improvement). The examples may be on any topic and in any area—scientific writing, popular-science writing, essays, technical reports or proposals, white papers, memos, letters, and blogs, to mention only a few. Show students what makes the paragraphs so effective. For example, note the following from Rachel Carson's (1962) *Silent Spring*:

> For the first time in the history of the world, every human being is now subjected to contact with dangerous chemicals, from the moment of conception until death. In the less than two decades of their use, the synthetic pesticides have been so thoroughly distributed throughout the animate and inanimate world that they occur virtually everywhere. They have been recovered from most of the major river systems and even from streams of groundwater flowing unseen through the earth. Residues of these chemicals linger in soil to which they may have been applied a dozen years before. They have entered and lodged in the bodies of fish, birds, reptiles, and domestic and wild animals so universally that scientists carrying on animal experiments find it almost impossible to locate subjects free from such contamination. They have been found in fish in remote mountain lakes, in earthworms burrowing in soil, in the eggs of birds—and man himself. For these chemicals are now stored in the bodies of the vast majority of human beings, regardless of age. They occur in the mother's milk, and probably in the tissues of the unborn child. (15–16)

Students can easily identify the opening sentence as the topic sentence (the order or pattern is general to specific), the focus on one topic or unity (our exposure to dangerous chemicals), the four brief supporting examples or details (completeness), and the flow from one example to another enhanced by the repetition of "they" (coherence). Ask them to find examples from a variety of sources and bring them to class for discussion. The more students can review the strengths of well-written paragraphs and the faults of poorly written paragraphs, the better.

However, we can do more than just cover the basics concerning paragraphs. In my experience, students need much more instruction on and practice with the concepts of coherence and cohesion and how both help to achieve flow, for example. We can provide them sample paragraphs lacking both qualities and ask them to revise as necessary. Then we can challenge them to create their own examples and revisions. We can ask them to experiment with different paragraph patterns (for instance, narrative, descriptive, question to answer, enumerative, analogous, specific to general) while emphasizing the benefits of beginning paragraphs in technical documents with a topic sentence when appropriate. We can ask them to experiment with paragraph length as well, helping them determine when a paragraph is complete and when a paragraph is not, or helping them understand why paragraphs in instructions are often no more than a sentence while paragraphs in reports are typically much more detailed.

## TEACHING THE MORE ADVANCED CONCERNS OF STYLE

Once students have some grounding in the basics of style, they find discussions of style at a more advanced level even more interesting and challenging. Covering style at the more advanced level perhaps works best in a separate undergraduate technical communication course or a course on technical editing. However, if we can also find ways to give at least some attention to the more complex elements of style in the introductory-level course, students will benefit.

Discussing more advanced issues about style and audience is a good place to start. At the basic level, we emphasize determining the reader's knowledge of the subject, motivations for reading, and style expectations and adjusting the style to the rhetorical context. At a more advanced level, we can explore in depth the concept of discourse communities and the complexities of these communities. For example, the discourse communities of civil engineers, industrial engineers, electrical engineers, and aeronautical engineers may have much in common, but there are also essential differences in these communities or domains. One possible assignment can be to require students to research one or more discourse communities related to their major by accessing O*Net at https://www.onetonline.org/ and the *Occupational Outlook Handbook* available online at http://www.bls.gov/ooh/. Both resources provide invaluable details on a wide variety of professions. To explore style more thoroughly, ask students to summarize what these resources might tell them about the communication values and the stylistic preferences of the discourse communities they research. For my introductory course, I have combined this exercise with an interview of technical professionals in each discipline (or different disciplines if students cannot find professionals in their field). Of course, an interview of a technical professional is common assignment in the introductory course, but I require students to focus on asking how these professionals must adapt their communications (written or verbal) for different audiences and what kinds of revisions they make when they do so. Students create their own questions, but they are encouraged to ask, for example, about the kinds of communications and presentations the technical professional must typically write or give, for what type of audience, and how and why they must adjust their prose for less expert audiences. (Beyond the scope of style, I also ask students to inquire about the many different software tools these professionals use for writing, presenting, video conferencing, and more, and I suggest students ask these professionals how much time each day they spend reading the communications of others, writing, presenting, and interviewing others for information. Students are often surprised to learn how much time professionals devote to communication during the average workday.) This interview assignment can teach students a great deal about the values of understanding and applying style choices in workplace communications. And, given the source (technical professionals), students find what they learn quite convincing.

At the more advanced level, we can also focus more in depth on plain language. Although I have taught an advanced course on technical writing style for over three decades, I recently developed a separate undergraduate course titled Technical Communication and Plain Language. I cover a wide variety of topics, including efforts to define or describe plain language and the history of the plain style, the elements and benefits of plain language, efforts to require plain language in government in our country and elsewhere, strategies for achieving plain language, the writing of low-literacy plain language, the basics of using plain language for page layout and for text in visuals, and efforts to professionalize plain language. The literature on these topics by those who teach technical communication continues to grow; scholars in this area include Beth Mazur (2000), Emily Thrush (2001), Katherine Miles and Jacqueline Cottle (Miles and Cottle 2011), Natasha Jones, Justin McDavid, Katie Derthick, Randy Dowell, and Jan Spyridakis (Jones at al. 2012), and Russell Willerton (2015). Exercises or brief assignments requiring students to research and discuss the origins of the plain style, or attempts to legislate a plain style here in this country, or plain language and international audiences, or various company and government guides to a plain style, can be worthwhile for teaching students about public debates on language or how a great deal of technical communication can be effectively written in a plain style.

Of course, students also benefit from knowing that not every subject for every audience can be distilled into a plain style. A complex style is often necessary and a better alternative. Demonstrate to students how a complex style may have longer sentences, more complex sentence structures, undefined jargon, longer paragraphs, and more challenging details but still be well written. For example, selected works of Lewis Thomas (1976) help demonstrate the differences between a plain style and a complex style. Many of the essays in his popular books written for a nonexpert audience use plain language, for the most part, while many of his professional publications in his field of cellular immunology reveal a greater complexity while also demonstrating the qualities of carefully chosen words, well-constructed sentences, and effective paragraphs. Requiring students to find and compare two examples of Thomas's writing on the same topic, one written mostly in a plain style and one written in a complex style, is a useful assignment. Or asking students to compare a paragraph by Thomas with a paragraph on a similar or different topic by another popular science author—for example, Stephen Jay Gould or Oliver Sacks—has its advantages.

Consider the following paragraph by Thomas (1976):

> It is probably true that symbiotic relationships between bacteria and their metazoan hosts are much more common in nature than infectious disease, although I cannot prove this. If you count up all the indispensable microbes that live in various intestinal tracts, supplying essential nutrients or providing enzymes for the breakdown of otherwise indigestible food, and add all the peculiar bacterial aggregates that live like necessary organs in the tissues of many insects plus all the bacterial symbionts engaged in nitrogen fixation in collaboration with legumes, the total mass of symbiotic microbial life is overwhelming. For sheer numbers, nothing can match the vast population of bacteria—or, at any rate, the lineal descendants of bacteria—which, taking the form of mitochondria . . . became essential symbionts for the production of oxidative energy in the cytoplasm of all nucleated cells, including our own, and in even greater numbers, the photosynthetic bacteria which became the chloroplasts of plants, without which neither plant nor animal life could ever have existed on the earth. Alongside, the list of important bacterial infections of human begins seems a relative handful. (216)

Ask students to discuss how the paragraph demonstrates the qualities of a complex prose style instead of a plain style. Students typically can readily identify the opening sentence as the topic sentence. They can easily identify the jargon or terms that would be unfamiliar to a novice. They can point out the examples—microbes in intestinal tracts, bacterial aggregates, and bacterial symbionts. And they recognize that two long sentences make up much of the paragraph. Considering Thomas is writing for his peers, the paragraph is effective and well written. It demonstrates a complex style where a simpler or plain style would be less appropriate.

We can also focus on many other styles in addition to a plain style and a complex style. Many students have never previously learned about other kinds of possible prose styles—for example, a telegraphic style, a verb style, a noun style, an affected or pretentious style, and various literary styles. John Fielden (1982) provides yet other styles: a forceful style, a passive style, a personal style, an impersonal style, a colorful style, and a less colorful style. Providing examples of each and discussing these examples, as well as requiring students to find examples for class discussion, can quickly show them the importance of style for particular audiences. Focusing in depth on pretentious styles provides numerous opportunities for students to experiment with style. Excellent resources are available for introducing students to pretentious prose, including Edward Tenner (1986), William Lutz (1989, 1996, 1999), John Barry (1991), and Tom Fahey (1990). Building on the knowledge provided by one or more of these resources (or knowledge provided by a brief class lecture), students can find published examples online and then discuss revisions they would make. Case studies and scenarios also work well with this topic.

We can focus more on a prose style for which the major fault is simply unintentional ambiguity due to poor writing. I have used case studies concerning the Three Mile Island leak, the space shuttle *Challenger* disaster, the *Exxon Valdez* oil spill, and the Ruby Ridge incident to demonstrate the unexpected and damaging consequences of poorly written policies and procedures, memos, directives, and more. Other major events—hurricane Katrina, the BP oil disaster in the gulf, and many more—typically are followed by local, state, or federal reports, and these reports often conflict with documents (e-mails, memos, reports) published by the companies involved. Students in introductory or advanced courses appear to enjoy equally the exploration and discussion of these issues. In some of my advanced courses, I have provided students the option of creating an original case study—with the necessary background and context, the specific confusing or otherwise poorly written language in one or more documents, and discussion questions.

Students in any level of technical communication course—from the undergraduate introductory course to a graduate-level course—can benefit from focusing in depth on the complex relationship between style and tone (and certainly beyond the correspondence assignment discussed earlier concerning a complaint and responses to the complaint). I view tone as more than the writers' attitudes toward the subject and the reader but also as a matter of writers' attitudes toward themselves. Writers explore a wide range of tones concerning their subjects—from flippant to serious and professional, from contempt to utmost respect. And writers do the same concerning their readers and regarding themselves. For example, they can use self-deprecation for effect or self-pity or self-loathing, or they can plead ignorance or convey arrogance or convey a lack of self-awareness. Or they can convey self-confidence, credibility, or authority. Ethos and pathos can provide great advantages for those who know how to manipulate tone for effect for both sincere and insincere purposes. Few topics also lend themselves better to teaching style than asking students to experiment with tone.

Provide students examples of different tones in paragraphs and ask them to identify the differences. Note how the first paragraph below uses a serious tone and a more formal style for the subject.

> A third line of research involves the human immune system itself, the primary victim of the AIDS virus. Actually, most if not all patients with AIDS die from other kinds of infection, not because of any direct lethal action of the virus itself. The process is a subtle one, more like endgame in chess. What the virus does, selectively and with exquisite precision, is to take out the population of lymphocytes that carry responsibilities for defending the body against all sorts of microbes in the world outside, most of which are harmless to normal humans. In a sense, the patients are not dying because of HIV, they are being killed by great numbers of other bacteria and viruses that can now swarm into a defenseless host. Research is needed to gain a deeper understanding of the biology of the immune cells, in the hope of preserving them or replacing them by transplantation of normal immune cells. This may be necessary even if we are successful in finding drugs to destroy the virus itself; by the time this has been accomplished in some patients, it may be that the immune system has already been wiped out, and the only open course will be to replace these cells. (Thomas 1992, 54)

Note the much lighter tone and more informal style in the following paragraphs:

> Warts are wonderful structures. They appear overnight on any part of the skin, like mushrooms on a damp lawn, full grown and splendid in the complexity of their architecture. Viewed in stained sections under a microscope, they are the most specialized of cellular arrangements, constructed as though for a purpose. They sit there like turreted mounds of dense impenetrable horn, impregnable, designed for defense against the world outside.
>
> In a certain sense, warts are both useful and essential, but not for us. As it turns out, the exuberant cells of a wart are the elaborate reproductive apparatus of a virus.
>
> You might have thought from the looks of it that the cells infected by the wart virus were using this response as a ponderous way of defending themselves against the virus, maybe even a way of becoming more distasteful, but it is not so. The wart is what the virus truly wants; it can flourish only in cells undergoing precisely this kind of overgrowth. It is not a

defense at all; it is an overwhelming welcome, an enthusiastic accommodation meeting the needs of more and more virus.

The strangest thing about warts is that they tend to go away. Fully grown, nothing in the body has so much the look of toughness and permanence as a wart, and yet, inextricably and often very abruptly, they come to the end of their lives and vanish without a trace. (Thomas 1980, 61)

Thomas's two examples provide for a good discussion not only on differences in tone and formality but also on differences in sentence and paragraph length, the use of similes and metaphors, and the author's respect for his two subjects.

Teaching students the need to recognize and to communicate within a hierarchy of authority or power can prove invaluable to their careers, too, and I see the matter as inseparable from the issue of tone. Of course, teaching students how to navigate their way through corporate politics is primarily a matter of professionalism, but tone is central here, too. Put another way, the tone deaf are often the least professional. Students should be taught why they must respond promptly and professionally to those in positions of authority or power. Students need to know when it will suffice simply to acknowledge the message they received or to provide what is requested or some other kind of response. Case studies, scenarios, and role playing can work well here. For example, we can ask students to share examples from their work experience in which paying attention to the power structure made a difference or in which ignoring the power structure led to negative consequences.

Humor also has a role in technical communication, particularly in documents and presentations for novices, and although many other topics perhaps deserve more attention in the introductory course, focusing some attention on how to achieve a humorous tone can be worthwhile. Of course, determining when using humor is appropriate in technical communication is a matter of the subject, purpose, genre, and audience. So discussions of humor provide a good opportunity to explore the rhetorical situation in more detail. In software documentation, I remind my students that humor works best in tutorials aimed at novices, whether the users are, for example, senior citizens, a volunteer or charity group, or a youth group. I point out that the *Dummies* series and *The Complete Idiot's Guide* series have been quite successful, both selling millions of copies over the past few decades or so. A stylistic analysis of how the humor is achieved in a document aimed at novices can be helpful. My students have chosen selections from one or more humorous documents aimed at novices and have critiqued the rhetorical and stylistic devices employed by the authors.

Some attention should be given to style and bias, too. Students in my advanced class on style have enjoyed discussions of gender bias in particular, with many admitting they were unaware of the many possibilities for bias in this area. We also explore racial bias, religious bias, scientific bias, political bias, cultural bias, corporate bias, and disabilities bias. I provide opportunities for students to explore these broad topics in depth in a research paper. Students need to know how to avoid sexist language, be aware of the challenges in addressing multicultural audiences, and know about corporate culture, corporate branding, and how to meet their employer's explicit as well as implicit expectations for professional behavior.

One productive exercise concerning bias is to require students to review the discussion about avoiding bias in the area of "APA Stylistics" on the Purdue Online Writing Lab website. See https://owl.english .purdue.edu/owl/resource/560/14/. Clicking on the links provided here for removing bias in language concerning race and ethnicity and sexuality reveals a wide variety of useful reports on these topics. Requiring students to summarize a report in a class discussion helps lend more authority to the seriousness of these issues. These issues can also be taught with case studies, scenarios, and role playing.

Finally, style and ethics can be addressed as well, and in a variety of creative ways. Increasingly, introductory technical communication textbooks include separate chapters on ethics, covering, for example, professional obligations, codes of conduct, and recognizing various kinds of unethical communication such as plagiarism and deliberately ambiguous language. Stylistic analysis of deliberately imprecise or ambiguous language can be revealing to students. For example, students can be asked to find and analyze an effective selection of propaganda or manipulative advertising to determine how the selection was constructed and to discuss why it might be successful for the intended audience. Opportunities for teaching students to experiment with language on this topic involve asking students to create scenarios or case studies concerning ethical issues they may confront in the workplace as technical communicators.

## CONCLUSION

We are in a better position now than we have ever been for focusing more attention on style in our technical communication courses. The literature supports our renewed interest in and the value for students of knowing and employing a wide range of strategies. Also, recent publications demonstrate that the interest in style remains strong, not just for academics or professionals or students but for the general public, whether these publications are new editions or reprints of classics such as Williams (2013), William Zinsser (2006) and Thomas and Turner (2011) or new works including Virginia Tufte (2006), Helen Sword (2012), and Stephen Pinker (2014). Providing students an awareness of this work can prepare them to handle many complex communication challenges they will face throughout their careers. They will acquire a better sense of when and how to be clear, concise, courteous, and persuasive. They will be better communicators because they will know many strategies they can employ for the best effect, whether they are writing, conversing, or presenting.

## DISCUSSION QUESTIONS

1.  In what ways can you use the prose style of some of your favorite authors, or the lyrics of some of your favorite songs, or the presentations of some of your favorite speakers (using YouTube or TED talks, for example) to demonstrate to your students the power or effectiveness of a particular style?

2.  Why should we focus not only on the essentials or basics of style but also on more advanced topics and issues? What are some specific ways we can benefit from focusing on more advanced elements, too?

3.  In what ways can creating scenarios, case studies, and role-playing scenarios help us understand the challenges for creating an effective prose style?

4.  In what ways can studying actual case studies on style choices help us understand the challenges of style?

5.  Why, ultimately, does possessing a good knowledge of style matter?

*References*

Anderson, Paul. 2014. *Technical Communication: A Reader-centered Approach.* 8th ed. Boston, MA: Cengage Learning.

Barry, John. 1991. *Technobabble.* Cambridge: MIT Press.

Bernhardt, Stephen. 1994. "Teaching English: Style in Writing." *The Clearing House* 67 (4): 181–82.

Blackburne, Brian. 2014. "Overcoming Workplace Writing Norms: Empowering Technical-Writing Students through Stylistic Analysis." *Programmatic Perspectives* 6 (2): 81–112.

Bradway, Becky. 1997. "Style Is Not Irrelevant: Finding Its Place in the Nonfiction Classroom." *Writing on the Edge* 8 (2): 21–31.

Carson, Rachel. 1962. *Silent Spring.* Boston, MA: Houghton Mifflin.

Dragga, Sam, and Elizabeth Tebeaux. 2015. *The Essentials of Technical Communication.* 3rd ed. New York: Oxford University Press.

Fahey, Tom. 1990. *The Joys of Jargon*. New York: Barron's.

Fielden, John. 1982. "'What Do You Mean You Don't Like My Style?'" *Harvard Business Review* 60 (3): 128–38.

Gerson, Steven, and Sharon Gerson. 2013. *Technical Communication: Process and Product*. 8th ed. New York: Longman.

Houp, Kenneth, and Tom Pearsall. 1968. *Reporting Technical Information*. Beverly Hills, CA: Glencoe.

Hudson, Randolph. 1961. "Teaching Technical Writing." *College Composition and Communication* 12 (4): 208–12.

Jones, Natasha, Justin McDavid, Katie Derthick, Randy Dowell, and Jan Spyridakis. 2012. "Plain Language in Environmental Policy Documents: An Assessment of Reader Comprehension and Perceptions." *Journal of Technical Writing and Communication* 42 (4): 331–71.

Lanham, Richard. 1974. *Style: An Anti-textbook*. New Haven, CT: Yale University Press.

Lanham, Richard. 1983. *Analyzing Prose*. New York: Scribner's.

Lanham, Richard. 2000. *Revising Prose*. 4th ed. Needham Heights, MA: Allyn and Bacon.

Lannon, John, and Laura Gurak. 2013. *Technical Communication*. 13th ed. New York: Longman.

Lutz, William. 1989. *Double-Speak: From "Revenue Enhancement" to "Terminal Living": How Government, Business, Advertisers, and Others Use Language to Deceive You*. New York: Harper and Row.

Lutz, William. 1996. *The New Doublespeak: Why No One Knows What Anyone's Saying Anymore*. New York: HarperCollins.

Lutz, William. 1999. *Doublespeak Defined*. New York: HarperCollins.

Markel, Mike. 2012. *Technical Communication*. 10th ed. New York: Bedford/St. Martin's.

Mazur, Beth. 2000. "Revisiting Plain Language." *Technical Communication* 47 (2): 205–11.

Miles, Katherine, and Jacqueline Cottle. 2011. "Beyond Plain Language: A Learner-Centered Approach to Pattern Jury Instructions." *Technical Communication Quarterly* 20 (1): 92–112.

Pinker, Stephen. 2014. *The Sense of Style: The Thinking Person's Guide to Writing in the 21st Century*. New York: Viking.

Sherman, Theodore, and Simon Johnson. 1975. *Modern Technical Writing*. 3rd ed. Englewood Cliffs, NJ: Prentice-Hall.

Stratton, Charles. 1979. "Analyzing Technical Style." *Technical Communication* 26 (3): 4–8.

Sword, Helen. 2012. *Stylish Academic Writing*. Boston, MA: Harvard University Press.

Tenner, Edward. 1986. *Tech Speak or How to Talk High Tech*. New York: Crown.

Thomas, Francis-Noel, and Mark Turner. 2011. *Clear and Simple as the Truth: Writing Classic Prose*, 2nd ed. Princeton, NJ: Princeton University Press.

Thomas, Lewis. 1976. "A Meliorist View of Disease and Dying." *Journal of Medicine and Philosophy* 1 (3): 212–21.

Thomas, Lewis. 1980. *The Medusa and the Snail: More Notes of a Biology Watcher*. New York: Bantam.

Thomas, Lewis. 1992. *The Fragile Species*. New York: Collier.

Thrush, Emily. 2001. "Plain English? A Study of Plain English Vocabulary and International Audiences." *Technical Communication* 48 (3): 289–96.

Tufte, Virginia. 2006. *Artful Sentences: Syntax as Style*. Cheshire, CT: Graphics.

Willerton, Russell. 2015. *Plain Language and Ethical Action: A Dialogical Approach to Ethical Content in the 21st Century*. New York: Routledge.

Williams, Joseph. 2013. *Style: Lessons in Clarity and Grace*. 11th ed. New York: Longman.

Zinsser, William. 2006. *On Writing Well: Thirtieth Anniversary Edition*. New York: Harper Perennial.

# 4

# TEACHING CONTENT STRATEGY IN PROFESSIONAL AND TECHNICAL COMMUNICATION

Dave Clark

Amanda, like all experienced instructors of professional and technical communication, has long engaged in the heavy lifting needed to help students keep up on the newest developments in the field, including toolsets and frameworks they will certainly encounter (e.g., InDesign, component content management) and theories and methodologies they need to know (genre theory, agile/scrum methodologies). In addition, she and her colleagues help students become familiar with the methods and foibles of their most likely workplace collaborators: project managers, business analysts, and subject-matter experts.

There's always something new. Professional and technical communication has always been an ambitious, expansive field that incorporates new approaches and puts its own shape on them. When knowledge management was at its peak in the early 2000s, Amanda and her students discussed Corey Wick's (2000) piece urging technical communicators to engage with knowledge management, along with key texts from *Harvard Business Review* and John Seely Brown and Paul Duguid (Brown and Duguid 2000). She also taught them the basics of content management and single sourcing. As *component content management* became the term of art, Amanda's students dug into Ann Rockley and Charles Cooper (Rockley and Cooper 2012), JoAnn Hackos (2006), and others who championed the move to topic-based authoring, XML, and DITA. These students graduated ready to help workplace teams transition to content management approaches that saved money (particularly on translation) while also improving clarity and consistency.

Now, in 2016, Amanda's reading of industry blogs and listservs, along with her attendance at recent conferences, has left her feeling she must again refresh her approach to advanced professional and technical communication. She must incorporate content strategy, a movement that

DOI: 10.7330/9781607326809.c004

asks technical communicators to look beyond producing ever-cheaper technical documentation that optimizes reuse to a more complete integration of their work with that of others across the enterprise (e.g., marketing, corporate communication, and training). This line of thinking isn't new: Rockley, Pamela Kostur, and Steve Manning argued for breaking down these organizational silos back in 2002 (Rockley, Kostur, and Manning 2002). But new tools and organizational priorities have made desiloing more practical and urgent than ever, and technical and professional communicators have a unique opportunity to take a leadership role in these changes.

But on a random Tuesday in October, where should Amanda start? Content strategy is still so new that definitions are in flux: it was just four years ago that Scott Abel and Rahel Anne Bailie's (Abel and Bailie 2014) *The Language of Content Strategy* attempted to codify content strategy terminology, and most articles and postings still begin by defining key terms. The first step, then, in any attempt to help students wrap their heads around the possibilities of content strategy, is to situate content strategy within the larger discourse. In what follows, I suggest how Amanda can provide students with some principles, guidelines, and ideas that can help them begin to grasp content strategy. I close with a call for the sharing of ideas and best practices that can help our classroom work begin to provide the guidance many students need as they enter contemporary technical communication workplaces.

## ASSIGNING FOUNDATIONAL TEXTS

In pulling together her initial lesson plans on content strategy, Amanda quickly discovers there has been an imbalance in content strategy discussions. Until very recently, content strategy was confined almost solely to nonacademic venues. A 2013 survey of TC practitioners rated content strategy as the single most important development in the field (Andersen et al. 2013); despite the widespread currency of content strategy in industry since around 2009 (Andersen and Batova 2015), very few academics ranked content strategy as having any importance. And while there has been some useful work specific to the teaching of content management (Evia, Sharp, and Pérez-Quiñones 2016; McShane 2008; Robidoux 2008), very little has emerged on how content strategy, which requires a broader understanding of organizational goals and processes than does content management, should best be taught to future communicators.

Fortunately, for those just getting started with content strategy, there are ample resources available from practitioners, and academics are beginning to provide alternative, peer-reviewed perspectives (cf. Andersen 2014; Hart-Davidson 2009), including two new edited volumes of *IEEE Transactions on Professional Communication*, both of which will be available by the time this volume is published. Amanda and her students will find the following texts particularly useful in learning what content strategy is, what it isn't, and how it might require some rethinking of the relationship of technical and professional communication to other fields:

1.  Ann Rockley and Charles Cooper (Rockley and Cooper 2012): *Managing Enterprise Content: A Unified Content Strategy.* Nine years after the original volume, this second edition provides key updates, including new definitions of content strategy and a focus on the now-critical concept of intelligent content.

2.  Kristina Halvorson and Melissa Rach (Halvorson and Rach 2012): *Content Strategy for the Web.* Because it focuses heavily on marketing and web-based communication, Halvorson and Rach's work provides a critical counterpart to the TC-focused work highlighted in the other recommended works.

3.  Scott Abel and Rahel Anne Bailie (Abel and Bailie 2014): *The Language of Content Strategy.* This book will make more sense after a reading of Rockley and Cooper, which provides an essential conceptual backdrop that helps ground the definitions in this text.

4.  Rebekka Andersen and Tatiana Batova (Andersen and Batova 2015): "Component Content Management: An Integrative Literature Review." This article provides a peer-reviewed, academic, comprehensive synopsis of all the relevant literature on component content management; in the process, it includes a very useful synopsis that contextualizes content strategy.

5.  Dave Clark (2016): "Content Strategy: An Integrative Literature Review." In my piece, I build on Andersen and Batova's work by providing a comprehensive synopsis of content strategy as an emerging method; in particular, I explore and differentiate key definitions of content and content strategy.

## DEFINING CONTENT STRATEGY

So what *is* content strategy? As Amanda's students will discover, it is difficult to provide a simple answer. There are multiple definitions, further complicated by the fact that definitions from web-intensive marketing strategists have a different emphasis than do those from technical communicators. As Abel (2013) suggests, "We lack a common understanding of the term. Content strategy means different things to different people.

And the differences in definition often are a matter of vantage point. It is not uncommon for those who come from the technical communication world to think of content strategy differently than those who hail from marketing, PR, user experience, information architecture, or mobile interaction design. As a result, confusion abounds" (14).

It is just this confusion that led to the production of *The Language of Content Strategy*, which defines content strategy as "the analysis and planning to develop a repeatable system that governs the management of content throughout the entire content lifecycle" (Abel and Bailie 2014). This definition is helpful in that it gives a sense of *scale* to the content strategy project (very large) and in that it emphasizes process and life cycle over any particular technology. Beyond that, it's a pretty abstract definition, and it's difficult to know what is included and what is left out.

It is helpful to consider some additional definitions to get a more complete picture of what is involved in content strategy. Ginny Redish (2012) is speaking of web-only content strategy in this quotation, but it nonetheless nicely captures what is new and unique about content strategy: "Following good practice in clear writing, let's turn the two nouns into a verb phrase: Content strategy = thinking strategically about your content. Thinking strategically means that instead of letting everyone post whatever content they want when they want with whatever messages they want, all the content on your website is part of your overall business plan" (37).

The final clause here is critical. Content strategy, at its heart, is about connecting formerly ad hoc or isolated departmental processes to the overall business plan of an organization. As such, content strategy is a unique opportunity for technical communicators, marketers, and PR specialists to better anchor their activities to the bottom line of the business. This kind of business relevance has been a goal for technical communicators for decades.

## CONTENT STRATEGY: LINKING CONTENT TO ORGANIZATIONAL GOALS

And there are still significant opportunities for content specialists to help develop new efficiencies and business relevance for professionals who write. Technical communication (and marketing and publishing) instructors must prepare their students for a working world in which they will be expected to work within and contribute to a content strategy. It's not a change that will necessarily come easy to those of us who, like Amanda, have a long history in technical and professional communication. As a field, we have been in transition from traditional, book-style publishing for at least two decades; the software startups I worked in during the late 90s were moving aggressively into single sourcing their documentation in order to optimize potential reuse across different media, even though we still produced print documentation. Today, few technical communicators continue to produce print content, but multimodal publishing is more critical than ever as devices and contexts for writing proliferate, so communicators have continued to rely on new tools and develop new methods for producing and managing content. At the same time, technical communicators (like those in other fields) are under continual pressure to do more with less while simultaneously better justifying their contributions in terms of the organizational bottom line.

"Content management," which a few years ago was suggested as the technical implementation of "knowledge management" (Clark 2002), was the best answer we had in the 2000s. Content management itself came in a number of varieties (Clark 2008), including web content management and enterprise content management, and technical communicators particularly latched onto component content management (CCM). In CCM, writers use topic-based authoring, moving from the linear structure common in books to modular chunks that can be reused in other documents and modalities. Topic-based authoring in TC is often structured using a standard like the Darwin Information Typing Architecture (DITA) and implemented using a proprietary or off-the-shelf content management system (CMS) that stores topics in an open standard like XML.

Properly tagged with metadata, topics can be easily rediscovered, reused, and repurposed, all without additional writing or revision. In fact, rewriting is actively avoided; a topic composed with reuse in mind can be used across multiple documents and can thus save significantly on translation costs. For example, a well-executed topic-based guide for installing a printer might reuse, say, 75 percent of its content in the manuals of an entire suite of printing products. Those topics, then, need not be retranslated each time, and writers and translators can focus solely on what is different among the printers.

Component content management has drawn interest from academic writers, who have published critiques of how it reconceptualizes rhetorical tasks (Bacha 2008; Clark 2002; Williams 2003), explorations of how it alters the field (Andersen 2011; Clark 2008), and explorations of the practitioner articles and research that suggest the potential changes we can expect in the careers of students (Andersen 2014; Hart-Davidson 2009). Meanwhile, workplace practices have evolved very quickly. As technical communicators, marketers, bloggers, journalists, and public-relations specialists have engaged with emerging tools and techniques that allow for better enterprise-wide repurposing, richer metrics for measuring content success, and intelligent content, the content management of just a few years ago is now thought of as one small piece of a larger movement toward a global content strategy that encompasses all phases of content life cycles and allows for a richer discussion of interactivity.

## MAKING CONTENT STRATEGY CONCRETE

Even with these definitions of what content strategy is and is not, it remains difficult to imagine what it is that a content strategist actually *does*, and that understanding is pretty important to being able to teach the methods and processes critical to the role. This diagram from Joe Gollner (2013), who blogs as the Content Philosopher, helps position content strategy appropriately in relation to other, better-defined activities (see fig. 4.1). Fundamentally, developing a strategy means systematically roping together acquisition, which incorporates the planning, designing, and creation of content; delivery, which includes the selection and assembly of content to meet user needs; and management, which includes database technologies and much of what we have always considered content management in technical communication. Finally, content engagement includes the development of feedback and social media mechanisms to involve stakeholders (readers) with the content.

In all cases, it is worth noting that content is, epistemologically, envisioned as a static thing to which management or engagement is applied, despite the emergence of intelligent content, which suggests autonomy but actually means content properly structured and tagged so as to be useful in a strategy that relies on reusing and adapting existing content to new contexts. The content isn't intelligent; it has a wrapping that allows us to use it intelligently.

But let's get even more pragmatic. If we hope to prepare students for content strategy work, we must have a solid grasp on what content strategists actually do. As I discussed earlier, there is precious little content strategy work from academics and/or in peer-reviewed publications, and much of what is out there (in industry publications, on blogs, and in popular-press books) is driven largely by establishing definitions and laying out the groundwork of the practice.

Still, examining some of the core texts of the field can be instructive in laying out some key principles, principles we must embrace in our teaching if we're going to properly prepare students for the world

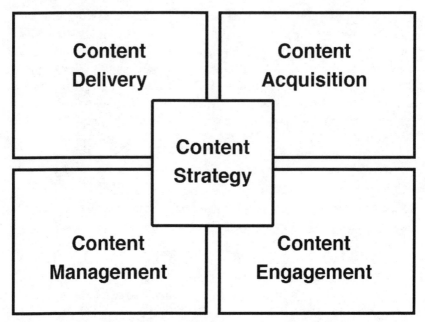

*Figure 4.1. Content Lifecycle Model (Gollner 2013)*

of content strategy. Many of the key texts seem remarkably similar in their approaches, which does provide some confidence that we're getting a reasonably comprehensive picture of the central activities. In assembling this overview of principles, I relied on Rahel Anne Bailie and Noz Urbina (Bailie and Urbina 2013), Sara Wachter-Boettcher (2012), Meghan Casey (2015), Kristina Halvorson and Melissa Rach (Halvorson and Rach 2012), Kevin Nichols (2015), and Rockley and Cooper (2012).

- **Content must be concretely aligned with business requirements.** Content strategy disrupts the traditional model of content as an ancillary, late-in-the-process user manual or brochure. Content strategists work proactively to align their work with the larger business goals of their organizations. Early in any content strategy initiative, strategists conduct a content audit, a process in which they determine whether their existing content meets the needs of their many internal and external stakeholders; a thorough audit assesses all organizational content in order to ensure it is achieving what it should for the good of the enterprise and establishes metrics for measuring future success.
- **Content is more collaborative and inclusive than ever.** An important initial goal of any content strategy initiative is break down silos (like traditional department and job boundaries) that impair content sharing and lead to unnecessary duplications of effort. In content strategy,

the content life cycle is the responsibility of everyone in the firm, and we broaden our understanding of content to include not just text and visuals but also information architecture and project management. In addition, we expand the scope of content to include texts that may have previously escaped notice, like product packaging, call-center support scripts, employee portals, and compliance information.

- **Content cycles and workflows must be carefully planned and implemented.** After strategists have identified all the producers, stakeholders, tools, and cycles, they must develop comprehensive methods for determining and modeling how content is produced, reused, managed, and delivered and must also manage the organizational change required to implement such a system. In doing so, they must strive to create *repeatable* systems that optimize efficiency and quality.
- **Content strategy requires a nontrivial array of new methods and tools.** Designers and users of a content strategy can often expect changes to the ways they write, store, share, and reuse content, both in terms of the methods and flows and in terms of the tools they are expected to use. Content strategists must have a good understanding of the available options and also must anticipate resistance and struggles for, say, writers required to move from a linear, hand-crafted model of content production to a topic-based authoring environment.

These content strategy basics tend to evoke one of two responses from technical writers. Experienced, senior writers often say, "I've always done all of this! I didn't know I was a content strategist." Less experienced technical writers ask, "Where is the writing?" My students most often fall into the second category. Many of our professional and technical writing students have switched into our program from another, less obviously pragmatic humanities program and are hoping to get a job in which they can simply write and edit as they have been trained throughout their schooling: individually, and without the pesky ties to larger business processes.

To these students, content strategy is an unwelcome intruder, insisting they learn countless new tools and methods and radically overhauling their perceptions of their futures. Some students, particularly the type who plan to be business analysts or project managers, take to this kind of work immediately. From some other students, we can expect some resistance. At least part of the task before us, then, is to persuade students of the importance of learning about content strategy. While all students who have graduated into roles in technical writing have learned XML, DITA, topic-based authoring, and other techniques on the job, if they are not given a bird's-eye view of the overall content strategy of their organization, they will be greatly delayed in their ability to contribute to company efforts at a higher level.

**ASSIGNMENTS**

Amanda and others new to content strategy should begin with assignments that, at least initially, are conventional in their writing and design, as students struggle to understand new concepts. At first, the goal should be establishing the importance of and demystifying content strategy, making it less abstract than it can seem after simply reading one of the many textbooks.

## The Applied Literature Review

To establish the importance of content strategy, bringing in industry guest speakers can be useful and eye opening, particularly for students who have a more conventional liberal arts background. But what can be even more helpful is requiring students to explore just how different the mainstream practitioner discourse about content is from anything they will see in academic journals or textbooks. Asking students to conduct and report on the contemporary discussions of technical communication and marketing content can give them a strong sense of just how much they have to learn. Some possible topics for exploration:

### Value

How does content strategy impact discussions about how content specialists argue for the value of their work to their organizations? Ten to twenty years ago, technical communicators relied on metrics such as reduced call volume and more amorphous topics like customer satisfaction. What new options does content strategy make possible, and how might they shape the future of content in the enterprise?

### Sources

How does discourse about content take shape in different media? Examine published texts, listservs, blogs, advertisements, and conference programs and proceedings. How are the sources, genres, and authors of work on content strategy shaping the discourse, and what does that shape suggest about future needs and directions for the discussion?

### History

How has technical communication discourse about content shifted in the past ten years? Twenty years? What new organizations, thought leaders, and alliances have emerged? What kind of work is less privileged as a result? What do these changes suggest about the future place of technical communication as a profession?

*The Needs Assessment, Content Inventory, and Content Audit*

The available texts on content strategy offer substantial, helpful guidance on how communicators can best assess their information infrastructure as an initial step toward developing a content strategy. Students can select a small organization to study, such as a student organization or their home academic department, and then they should conduct and report on the full range of what is available. There are three key phases:

- **Needs Assessment.** It is impossible to perform a content audit without a detailed assessment of organizational and stakeholder needs. As Rockley and Cooper (2012) note, the key is determining what your customers (internal and external) really need. To pull together this information, students should interview stakeholders across functionalities; marketing, documentation, sales, and other segments of the enterprise perceive different needs. Rockley and Cooper offer an excellent guide to help students get started, with an emphasis on identifying pain points that suggest areas for improvement. Nichols (2015) offers a particularly helpful breakdown of questions related to business and project strategy goals.

- **Content Inventory.** The content *inventory* is, as Nichols suggests, a complete list of all of an organization's content and content processes; the intent is not to evaluate but to simply uncover all the different kinds of content available to and produced by the organization and how they are produced. Students should develop charts and follow best practices from one of the many sources (Nichols is a good one) on how to systematically present the complete list, broken down by ID, title, genre, description, functionality, purpose, metadata, and so forth.

- **Content Audit.** Now equipped with an understanding of organizational needs and a comprehensive list of content options, students can conduct an *audit*, which assesses the content *inventory* in light of stakeholder *needs*. Fundamentally, the purpose of the audit is to assess the quality, rhetorical appropriateness, tone, and so forth in terms of customer needs. What content is doing its work effectively? What content needs to be improved? What content doesn't yet exist and should be created? What content can be reused? What existing processes could be improved, streamlined, revised, eliminated?

*The Tool Training*

Not every student will be persuaded of the use of learning XML and structured authoring tools, and I don't advocate a technology-heavy approach to teaching content strategy. Tools change frequently, and the sorts of tools required for a corporate-style content strategy can be cost prohibitive to many universities (although it's worth noting that there are countless open-source alternatives to industry-standard tools like Arbortext and MadCap Flare). I strive, instead, for students to understand the range of options available to them and develop *conceptual* understandings of the potentials of topic-based authoring, XML, intelligent content, and data-driven reuse systems.

An effective method for assessing student understanding is a simple topic-based authoring project. Working with one problem or opportunity identified in their analysis of the content set in the previous project, students develop a paper-based prototype for a solution, a solution that includes identifying the relevant tools and methods and, when applicable, a breakdown of how those tools will be employed in resolving the identified issue. For such a project, there are many available prototyping tools (at this writing, Balsamiq and InVision are among the most popular) that allow students to create clickable mockups that can be shared and tested digitally, allowing for the rapid accumulation of testing data.

*The Content Strategy*
Having shown their understanding of auditing and tools (and having received substantial feedback on their work thus far), students are prepared to develop the initial steps toward an overall strategy for tackling the content ecosystem of their identified organization. It's important to note that a full content strategy involves many more components and much more complexity than would be possible to take on in a single report project or in a single semester, so our goal here cannot be a complete plan but can be a reasonable structure in which students can note strengths, weaknesses, and known omissions. The report they produce should, at minimum, include the following key elements, some produced earlier in the semester:

- before and after diagrams of the content ecosystem;
- the analysis of business and stakeholder needs;
- the content audit, identifying pain points and contradictions between stakeholder needs and existing content and content practices (the full content inventory should be included as an appendix);
- lists of content types (genres);
- workflow diagrams;
- notes toward the creation of a content model, tool infrastructure, style guides, and all additional materials needed for a complete rollout of the strategy.

## THE RHETORICAL REFLECTION
Finally, the students should be asked to reflect on their accomplishments. What are the strengths and limitations of their content strategy

plan, and what are the strengths and limitations in their own under-standings of how content strategy can work to improve an organization? How has the project made them think differently about their place in an organization's communication efforts? And, do they believe their content strategy could help an organization to better succeed?

These assignments, ultimately, are *reports* intended to challenge students cognitively in terms of their content and also challenge their burgeoning abilities as writers of technical content. Still, there is obviously a limit to how much students will gain from this arm's-length approach to content strategy, and an ideal approach might include a second semester of content strategy, or an internship in which students would get their hands dirty with more real content and encounter some of the real problems that crop up in the less ideal world outside their scenarios.

## CONCLUSION

There is no question that we need more research into content strategy; as I discussed in the introduction to this chapter, there has been virtually no empirical or peer-reviewed work on content strategy despite the existence of a massive and growing practitioner literature. There has been even less work into best practices for helping students make the conceptual and technological leaps from a more traditional understanding of text production into content strategy. And we need better bridging materials. Sarah O'Keefe and Alan Pringle's *Technical Writing 101* (O'Keefe and Pringle 2011), for example, is a frequently used text for introducing students to the profession of technical writing (as distinct from the many technical writing texts aimed at students who will be writing technical information as engineers or scientists), but it contains only a single chapter relating to topic-based authoring and XML. Their *Content Strategy 101* (O'Keefe and Pringle 2012) volume, on the other hand, like many content strategy texts, best fits the needs of current professionals rather than students.

I have found that the incoming goals of students are different than they were even a few years ago. I have seen a distinct downturn in the number of students in professional and technical writing who hope to edit mystery novels or work in the magazine industry; the message that these fields are troubled and offer few opportunities seems to be getting through. And increasingly, students are placing into a wider array of job types than ever before; our most recent graduates are in writing roles as diverse as corporate communications, public relations, project management, business analysis, and social media. Technical writing is not the clear, singular path it once was. Content strategy is, in some ways, the perfect evolution for contemporary practice in technical communication in that learning about content strategy gives students a broader and yet more nuanced understanding of organization-wide communication.

## DISCUSSION QUESTIONS

As you discuss the emergence, importance, and practicalities of content strategy with students, the following kinds of questions will be important to helping them situate it within the history and ongoing evolution of technical communication.

1. How does content strategy reconfigure the work of technical communicators? How has the day-to-day work of a technical communicator changed over the last twenty years? Over the last forty?

2. In what ways does content strategy enable greater alignment of technical communication goals with the goals of an enterprise? In what ways will technical communication continue to be challenged by the idea that it can be easily outsourced and/or that it's an afterthought in the larger mission of the organization?

3. In what ways does content strategy challenge the traditional work of technical communicators? Consider the growth of alternative organizations and the shrinking of the Society for Technical Communication. What new competencies are now required to be an effective technical communicator? Is technical communication even the same field?

4. What additional research do we need in order to understand the implications of content strategy for the larger workforce, let alone for technical communication?

### References

Abel, Scott. 2013. "The Importance of Vision in Content Strategy." *Intercom* 60 (5): 14–16.

Abel, Scott, and Rahel Anne Bailie. 2014. *The Language of Content Strategy*. Laguna Hills, CA: XML.

Albers, Michael. 2003. "Single Sourcing and the Technical Communication Career Path." *Technical Communication* 50 (3): 335–44.

Andersen, Rebekka. 2011. "Component Content Management: Shaping the Discourse through Innovation Diffusion Research and Reciprocity." *Technical Communication Quarterly* 20 (4): 384–411.

Andersen, Rebekka. 2014. "Rhetorical Work in the Age of Content Management: Implications for the Field of Technical Communication." *Journal of Business and Technical Communication* 28 (2): 115–57.

Andersen, Rebekka, and Tatiana Batova. 2015. "The Current State of Component Content Management: An Integrative Literature Review." *IEEE Transactions on Professional Communication* 15 (3): 247–70.

Andersen, Rebekka, Sid Benavente, Dave Clark, William Hart-Davidson, Carolyn Rude, and Joann Hackos. 2013. "Open Research Questions for Academics and Industry Professionals: Results of a Survey." *Communication Design Quarterly* 1 (4).

Bacha, Jeffrey. 2008. "Single Sourcing and the Return to Positivism: The Threat of Plain-Style, Arhetorical Technical Communication Practices" In *Content Management: Bridging the Gap Between Theory and Practice*, edited by George Pullman and Baotong Gu, 143–59. Amityville, NY: Baywood.

Bailie, Rahel, and Noz Urbina. 2013. *Content Strategy: Connecting the Dots between Business, Brand, and Benefits*. Laguna Hills, CA: XML.

Brown, John Seely, and Paul Duguid. 2000. *The Social Life of Information*. Boston, MA: Harvard Business School Press.

Casey, Meghan. 2015. *The Content Strategy Toolkit: Methods, Guidelines, and Templates for Getting Content Right*. Berkeley, CA: Peachpit.

Clark, D. 2002. "Rhetoric of Present Single-Sourcing Methodologies." *Proceedings of the 20th Annual International Conference on Computer Documentation, Toronto, ON*, 20–24. New York: ACM Press.

Clark, Dave. 2008. "Content Management and the Separation of Presentation and Content." *Technical Communication Quarterly* 17 (1): 35–60.

Clark, Dave. 2016. "Content Strategy: An Integrative Literature Review." *IEEE Transactions on Professional Communication* 59 (1): 7–23.

Gollner, Joe. 2013. "The Technology Side of Content Strategy." *Intercom* 60 (5): 17–20.

Evia, C., M. R. Sharp, and M. A. Pérez-Quiñones. 2016. "Teaching Structured Authoring and DITA through Rhetorical and Computational Thinking." *IEEE Transactions on Professional Communication* 58 (3): 328–43.

Hackos, JoAnn T. 2006. *Information Development: Managing your Documentation Project, Portfolio, and People*. 2nd ed. New York: Wiley.

Halvorson, Kristina, and Melissa Rach. 2012. *Content Strategy for the Web*. Berkeley, CA: New Riders.

Hart-Davidson, William. 2009. "Content Management: Beyond Single-Sourcing." In *Digital Literacy for Technical Communication: 21st Century Theory and Practice*, edited by Rachel Spilka, 128–44. New York: Routledge.

McShane, Becky Jo. 2008. "Why We Should Teach XML: An Argument for Technical Acuity." In *Content Management: Bridging the Gap Between Theory and Practice*, edited by George Pullman and Baotong Gu, 73–85. Amityville, NY: Baywood.

Nichols, Kevin P. 2015. *Enterprise Content Strategy: A Project Guide*. Laguna Hills, CA: XML.

O'Keefe, Sarah, and Alan Pringle. 2011. *Technical Writing 101*, 3rd ed. Durham, NC: Scriptorium.

O'Keefe, Sarah, and Alan Pringle. 2012. *Content Strategy 101: Transform Technical Content into a Business Asset*, 3rd ed. Durham, NC: Scriptorium.

Redish, Janice. 2012. *Letting Go of the Words: Writing Web Content That Works*. 2nd ed. Amsterdam: Elsevier/Morgan Kaufmann.

Robidoux, Charlotte. 2008. "Rhetorically Structured Content: Developing a Collaborative Single-Sourcing Curriculum." *Technical Communication Quarterly* 17 (1): 110–35.

Rockley, Ann, and Charles Cooper. 2012. *Managing Enterprise Content: A Unified Content Strategy*. 2nd ed. Berkeley, CA: New Riders.

Rockley, Ann, Pamela Kostur, and Steve Manning. 2002. *Managing Enterprise Content: A Unified Content Strategy*. Berkeley, CA: New Riders Press.

Wachter-Boettcher, Sara. 2012. *Content Everywhere: Strategy and Structure for Future-Ready Content*. Brooklyn, NY: Rosenfeld Media.

Wick, Corey. 2000. "Knowledge Management and Leadership Opportunities for Technical Communicators." *Technical Communication* 47 (2): 515–29.

Williams, Joe. 2003. "The Implications of Single Sourcing for Technical Communicators." *Technical Communication* 50 (3): 321–27.

# 5

# TEACHING GENRE IN PROFESSIONAL AND TECHNICAL COMMUNICATION

Brent Henze

Wendy, a new instructor, is introducing her students to the first writing assignment in her introductory technical writing class, a memo-style informational report in which students are expected to tabulate some sales data, then summarize and interpret that data for their employer. She's prepared a data set for students, along with some details about the hypothetical situation in which they're working: they're sales managers who have compiled the results of their department's completed sales over the past three months, and their employer has asked them to report on their sales as part of a regular quarterly review that'll be used to guide future decisions about marketing, staffing, and product development.

In this teaching scenario, Wendy is inviting her students to participate in a generic situation—that is, a situation requiring the use of a genre (the quarterly-sales-report genre). One traditional way to guide students' efforts would be to introduce the conventional characteristics of quarterly sales reports, to provide a model of one of these reports, and to highlight the genre rules of these texts: the expectations regarding document length, formality of style, manner of formatting data, document design characteristics, and so forth. Using these rules as well as the provided data set, students would then follow the rules of this reporting genre, producing a text that mimics what a real-world quarterly sales report might look like.

However, this traditional approach raises several complications, some very practical and obvious (is it even possible to specify the typical page length for such a document?) and some having to do with less obvious yet more fundamental problems involving how students learn to apply genre knowledge. In what follows, I'll highlight some of these problems, referring to the findings of contemporary genre research in professional and technical communication. I will also provide examples of how

10.7330/9781607326809.c005

Wendy might modify her approach so as to introduce this assignment and teach its elements more effectively.

Whether or not we explicitly talk with students about the principles of genre in our professional and technical writing classrooms, we as teachers should know how genres work. After all, according to our current theories of genre in professional and technical writing, genres are not optional aspects of writing, something we can take or leave. Genres are part of the fabric of the social action that occurs in any communication event; they enable and define the set of possibilities and even the roles available to writers in every communication context.

Wendy's understanding of genre will govern how she designs assignments, what she prioritizes in her classroom, and how she talks with her students about the writing process. Anis S. Bawarshi (2003) notes that traditional pedagogies have treated writing as a fundamentally individual act. A genre approach, by contrast, not only treats writing as a fundamentally social rather than individual process, but it even views writers themselves as fundamentally social insofar as writers' intentions and the details of their writing processes, from invention to delivery, are regulated by their rhetorical situations. To put it more simply, a genre approach asks us as teachers to pay attention to how the entire process of writing, not just the final delivery stage, is social (from problem definition to invention of arguments to presentation and delivery). This social orientation toward writing actually works great for teaching professional and technical writing since, even more than in other areas of writing, teachers of professional and technical communication are already focused on how professional writers work together in teams and organizations toward shared objectives.

## TWO STANDPOINTS ON GENRE

If you ask students to define the term *genre,* they're likely to say genres are nameable categories of the texts commonly encountered: mystery novels, epic poetry, science-fiction films. In addition to literary or film genres, students might be able to list common professional communication genres, such as reports, proposals, product manuals, business letters, e-mails, résumés, meeting minutes, instructions, press releases, impact statements, bids, service contracts, and product specifications. Texts in a genre can be described according to their standardized forms, styles, and contents. So, a business letter is a distinctive genre that can be specified by its form, content, and linguistic characteristics: it's from a few paragraphs to a few pages in length and deals with business-related topics; it is written in active voice and a relatively formal style, begins with a stylized greeting ("Dear So-and-so:"), and ends with a stylized closing ("Sincerely, Biff"); it's mailed from sender to recipient; and so forth.

This formal-linguistic view of genre has the advantage of familiarity: it's how we often recognize genres in the wild, and, in fact, it's a helpful way to name documents. But it doesn't help us to know *why* genres are the way they are, or how they came to be that way, or how the conventional forms of genres accomplish the work we mean to accomplish when we write. It also doesn't explain the great variation we find within genres—not all business letters look the same, nor do all mystery novels, environmental-impact statements, or service contracts. This definition encourages a mechanistic approach to teaching that mystifies genre-based writing and that, according to some researchers, doesn't work very well to prepare students to write when they enter the workplace.

To help students understand genres as tools for accomplishing practical goals, we must look elsewhere. In what follows, I introduce key insights from the scholarship in rhetorical genre studies, particularly where it applies to teaching and learning. I offer strategies for approaching genre in the professional and technical communication classroom and consider arguments over whether it is possible or desirable to teach genres directly. I offer suggestions for how to incorporate genre into your professional and technical communication course design to help students produce new texts and to analyze professional writing contexts more effectively. And I conclude with practical advice and sample activities that can be used in your classes.

## GENRE THEORY AND ITS RELATIONSHIP TO TEACHING AND LEARNING

In the last four decades, writing teachers and scholars have turned to sociorhetorical theories of genre to account for how genres actually work in professional workplaces and other communication environments. It's customary to begin any discussion about these developments by pointing to Carolyn Miller's (1984, 159) definition of genres as "typified rhetorical actions" that occur in "recurrent situations." Miller argues that "a rhetorically sound definition of genre must be centered not on the substance or the form of discourse but on the action it is used to accomplish" (151). Although she was not the first to highlight problems with formalist definitions of genre or to propose more situational, socially grounded definitions, Miller's 1984 elaboration of this shift launched a research trajectory that continues today. This perspective is often referred to as "North American genre studies" or "rhetorical genre studies" (RGS).

According to the rhetorical genre studies standpoint, the essence of genre is not the formal regularity of the genre's texts but the typified rhetorical situation: genres work because those who use them perceive them to be familiar responses to their rhetorical situations. Because the members of a discourse community share a common context, they recognize texts produced according to the community's habits (Bawarshi and Reiff 2010, 67). Genres are part of the "lay of the land" writers need to become familiar with when they enter a new workplace or other situation, just as they learn other communal standards, habits, or expectations.

Although all acts of communication are social (involving at least a sender and a receiver), genres are social in a more powerful way: they describe activities we perform according to shared understandings and purposes in social contexts that govern not just how we write but also what symbolic actions it's possible to perform. According to Charles Bazerman (1994), "In-so-far as they identify a repertoire of actions that may be taken in a set of circumstances, [genres] identify the possible intentions one may have" (82). That is, situations come with their own set of things it's possible do—our "intentions" are not fully our own but are constrained by our circumstances. Genres are a community's "repertoire of situationally appropriate responses to recurrent situations" (Berkenkotter and Huckin 1995, ix).

In this view, the rhetorical action is paramount, not the features of the resulting text. In fact, it's often helpful to talk about genres not as nouns but as verbs, shifting students' attention toward workplace *actions* rather than workplace *texts*. As a communicator in a workplace, I don't sit down one morning and decide to write a recommendation report. Rather, I might decide that I need to recommend something to my colleagues—an action that makes sense in my rhetorical context. Because I'm familiar with the activities and perspectives of my workplace, and because I've encountered similar situations before (as a reader or as a writer), I know something about how the professional act of recommending happens here: how there's a protocol, there are expectations about what counts as good reasons and how evidence is used, there's a process for passing recommendations through the workplace's hierarchy, and so forth. My intention is shaped by all these factors, and when I perform the rhetorical action of recommending in this context, what comes out is an instance of a genre, the recommendation report, an object others in my community will (if I acted effectively) recognize and know how to respond to.

*Models for Teaching Genre*

Despite the sophistication of these new theoretical models of genre and their enormous contributions to how we understand genres in use, genre's impact on professional and technical writing pedagogy has been both inconsistent and conflicted. For instance, even when contemporary textbooks in technical writing discuss genres as textual regularities emerging from disciplines' communicative actions, habits, and needs (in agreement with rhetorical genre studies), these same textbooks almost invariably introduce specific professional genres (e.g., proposals, instructions, business memoranda) as relatively fixed textual forms that can be modeled and practiced.

Although this formal textual approach to genres may simply be expedient, it still prioritizes the most stable, visible, and apparently important genres while deemphasizing the more ineffable communicative practices occurring in technical workplaces—all the local rhetorical work and social negotiations that add up to a professional community's genre work. Bawarshi and Mary Jo Reiff observe that "the challenge for RGS has been how to develop genre-based approaches to teaching writing that attend to this dynamic—how, that is, we can teach genres in ways that maintain their complexity and their status as more than just typified rhetorical features" (Bawarshi and Reiff 2010, 189).

In fact, several empirical studies have raised doubts about whether it is even possible to explicitly teach genres outside the authentic contexts of their use (Freedman 1993; Freedman, Adam, and Smart 1994), leaving professional and technical communication educators in a bind: if genres are fundamental to communicators' membership and performance in the workplace (and other discourse communities), but genres cannot be effectively taught outside those contexts, what role is there for genre study in the technical communication classroom?

The pedagogical literature offers two responses to this puzzle. First, some argue that explicit teaching of professional genres remains a realistic goal in the classroom but that we need better methods for doing it. Second, a variety of pedagogical approaches have been developed that focus not on teaching genres themselves but on cultivating students' *genre awareness* in an effort to prepare them to learn authentic genres as they enter the contexts of their use.

Amy Devitt, Richard Coe, and Charles Bazerman, among many others, have argued that student exposure to genres (and to RGS-informed theories of genre) has learning benefits beyond the direct goal of mastering particular genres. As Bawarshi and Reiff (2010) describe it, genres serve as "learning strategies or tools for accessing unfamiliar writing situations" (191). Devitt (2004) offers a pedagogy based upon "meta-awareness of genres, as learning strategies rather than static features" of text (197). In Devitt's genre-awareness model, students engage in a variety of critical-analytic activities: working with existing examples of a target genre, applying rhetorical-analysis techniques to imagine how the texts in a genre fulfill common purposes; repurposing texts to see how different rhetorical situations affect genre characteristics. By playing around with a genre's elements in a variety of ways, students gain perspective on how those elements work together to accomplish generic purposes in particular contexts (Devitt 2009).

Richard Coe (1994) offers a similar pedagogy, proposing two frames that can be used to help students approach genres: the *archeological* and the *ecological.* In the archeological approach, teachers ask their students to examine and experiment with genres as archeological artifacts, empirical evidence of the deeper structural, relational, and rhetorical contexts in which they operate, just as an archeologist might study a set of artifacts and then construct models of the lives and habits of the people who used those artifacts. When Wendy asks her students to read several different examples of quarterly sales reports and consider possible situation-specific causes for some of the variations among those examples, she's asking them to take an archeological approach.

So too, students can examine genres as parts of a sociorhetorical ecosystem, revealing that "genres are situated in contexts and need to be explained as somehow fitting those contexts, for genres evolved as people adapted to communicative situations and their contexts" (Coe 1994, 163). Just as natural ecosystems are composed of a variety of organisms that fill unique niches in relation to one another, in a genre ecosystem, the various kinds of communication (and other social interactions) work together, and the details of each genre can be partly explained by looking at how they relate to the rest of the ecosystem. By asking her students how the quarterly sales report fits into the workplace's larger system of communications and works in concert with other genres to achieve workplace goals, Wendy is highlighting the ecosystems model of genre.

Using both these frames, we can use genres as probative tools to illustrate how writing works in real-world contexts, helping students prepare for their own entry into those contexts. Since we want to help students see how genres work by studying genres in use (rather than merely to teach students how to use specific genres), Coe notes, students' efforts needn't even be entirely successful to have pedagogical value. He quotes John Swales, who observed that "the rationale behind particular genre features may prove elusive, but the process of seeking for it can be enlightening . . . for instructor and student" (quoted in Coe 1994, 161).

### Using Genre Concepts to Talk about Flexibility, Variation, and Choice in Writing

One of the most exciting implications of genre in professional and technical communication classrooms is that genre theory offers us a way to encourage mindful choice in students' writing. Contemporary theories of genre help explain how regularities arise, but these theories are equally interested in issues of difference and choice.

In a field like professional and technical communication, the range of technical and professional writing genres is tremendously varied in comparison to the apparent uniformity of the academic essay or paper. But that very variety can be intimidating for students. On the one hand, writing as a professional communicator means something different in each new circumstance; mastery over one familiar genre doesn't mean mastery over other genres. On the other hand, students might reasonably wonder, why don't *all* instances of a given genre look exactly the same? Why do writers feel their work is a process of making choices rather than simply following predetermined rules?

Genre theory opens up the vistas of writerly choice, which might seem a mixed blessing to students. We can help students by highlighting what choice means (it's not a free-for-all) and emphasizing how professional writers make choices in response to their circumstances. Genre gives us ways to talk with students about how writing situations *always* offer great opportunities for choice but also about how any choice we make is shaped and constrained by rhetorical context. It also encourages discussions about what it means to be creative, to defy expectations for strategic purposes, and even to assess the capabilities and limitations of particular genres, recognizing that writers make an important choice even by deciding to tackle the rhetorical situation using a particular genre.

Returning to Wendy's class: instead of introducing her students to workplace reports by showing them a *prototypical* workplace report (one that demonstrates all the "rules" of this genre), she might instead provide examples of a few different quarterly reports, each accomplishing a similar set of purposes but demonstrating a range of authorial choices (from organization to style to content) so her students can witness how workplace communicators make strategic choices that vary according to circumstances. This discovery helps explain many otherwise frustrating things about genres, such as why a rhetorical choice that succeeds in one setting fails in another, or why we find such surprising variability in apparently clear-cut genre situations. It also helps us understand why, as we move from context to context, or even from day to day, we are occasionally surprised to discover that the rules have changed and the genres we've come to depend upon don't fit as perfectly as we expected—an experience anyone who has changed jobs or even moved from one class to another will recognize.

As teachers, we want students to make choices in their writing, yet we also hold them accountable for the quality (even, sometimes, the correctness) of some of those choices; even if there's no single way to solve a problem, not all approaches are equally good. Genre theory helps explain how the writer's freedom and constraints go hand in hand. Susan Katz notes that genres are a combination of constraints and opportunities: they "constrain writers by limiting the form, style, language, and content that are appropriate in particular situations," yet these constraints only narrow the range of options, leaving a great deal of room for decision making (quoted in Artemeva 2006, 23). By eliminating the need to make certain choices, genres actually free up the writer to make finer decisions about what to do within those constraints; these decisions can be driven not by the genre's demands but by the demands of the specific circumstances at hand.

Because genres provide plenty of room for choice while also constraining certain kinds and degrees of choice, the most adept users of a genre turn out not to be writers who hew to the line most closely but those who make the most of the genre's flexibility, strategically deviating from convention or exploiting opportunities for choice without overstepping firm expectations. As Wendy and her students talk through several examples of business reports, she might pay special attention to incidents in which the writer has seemingly violated a norm but has strengthened the resulting text (for instance, by judicious use of informal tone, or by using an unorthodox structure). Experienced users come to recognize where they can deviate from genre expectations and how to bring readers along with them, whereas less skilled users are more likely to commit gaffes, or, conversely, to stick to the genre rules even when their particular rhetorical situation calls for a bolder choice.

So, although genres constrain and regulate how we write, they also enable innovation and encourage writers to respond to rhetorical circumstances creatively. Over time, students' increasing mastery of genres enables greater flexibility and freedom; the "power users" of a genre are versatile, strategic, and, yes, even creative. By emphasizing genre principles in our teaching, we can focus on helping students develop rhetorical savvy, judgment, and capacity in any workplace (or other) setting. In fact, as new technologies arise, as new workplace structures appear, and even as students move from one position to another, they'll need to learn genres continually. The principles of genre help us prepare students to engage in that continuous, adaptive, and situated learning.

## Interdependence of Genres

Genre scholars have also studied how genres can be related to one another, forming groupings theorists have used a variety of metaphors to articulate, including *genre systems* (Bazerman 1994; Berkenkotter 2001; Russell 1997; Yates and Orlikowski 2002), *genre sets* (Devitt 2004), *genre repertoires* (Orlikowski and Yates 1994), and *genre ecologies* (Spinuzzi 2003; Spinuzzi and Zachry 2000). These insights are especially valuable for teaching since they stress the value of introducing writing assignments as parts of larger sequences or networks of communication; a genre-based professional writing assignment will probably involve students writing documents that respond to previous documents and that prompt future communications. As I discuss later in this chapter, assignment sequences are a particularly useful strategy here since they not only give students experience with multiple genres but also encourage them to treat professional writing as part of an ongoing process of collective activity in the workplace.

As Mikhail Bakhtin (1987) explained, far from being "stand-alone" texts, all writing responds to prior communications, and often these sequences follow patterns of generic response (as, for instance, a Request for Proposals creates the exigence for a proposal in response, or a letter of inquiry calls for one or more genres in response: an informational letter, a refusal, or a service call). Genres can be seen as the set of "moves" available in a communication situation. If each genre exists as a patterned response to a recurring exigence in a familiar communication context, it's not surprising that each genre might intersect with other genres in that context in characteristic ways, filling "niches" in a genre "ecology" (Spinuzzi and Zachry 2000). Genres not only respond to previous genres and summon generic responses according to repeating patterns, but they also shoulder up against one another in ways we can ask students to trace, describe, and analyze.

*Teaching Students about the Relationship between*
*Textual Forms and Rhetorical Genres*

The genre principles outlined above encourage us, as teachers of writing, to spend much more time helping students become more receptive to rhetorical situations than we might if we considered genres to be merely forms or containers for text. We want students to see that all communications are *context specific*: arising in a specific place, at a specific moment, motivated by particular conditions. Savvy communicators don't begin by thinking about textual forms or rules; they begin by seeking to grasp the conversation any new message will be entering into: what it's about, who's involved, why it's important, and how they're positioned in it.

In practice, that means routinely prompting students to consider the situations when they're approaching a writing task: asking them to spend as much energy forming a clear picture of where their writing is going, and who'll be using it, and why, and when, as they spend actually developing content or arguments. If students hear us focusing on the wrinkles of actual rhetorical situations, puzzling over the what-ifs, they'll be more likely to pay attention to details of their rhetorical situations as well—and that's the key to shifting students' attention toward producing legitimate writing rather than merely aping a conventional form.

Let's return to the example of Wendy, who has introduced her students to the sales-report assignment. Instead of simply providing students with a set of report-writing rules or conventions and asking students to follow those rules to produce their reports, she takes a different approach, asking students why such a report might be used in a sales workplace. Why would a writer's supervisor ask for a report, as opposed to simply asking for a data set, every quarter? Why compose this information in written form (instead of asking for raw data tables, or asking for a verbal accounting of the quarter)? What functions or purposes other than mere transmittal of data might be achieved by a written report, and who in the workplace might benefit from that work?

Wendy could start her class by briefly "profiling" the people involved in writing or reading these reports, inviting students to contribute their thoughts about how the report might foster business purposes or build relationships among the people involved in the workplace. To put it another way, she would talk about the report as part of a larger process of *doing business* that involves not just writing and reading but also all the other relationship-building, management, and interactive efforts of people in this work environment. Rather than trying to simplify the writing task by treating the report as an object of individual work, she would encourage her students to complicate the report by showing how much it is a part of the larger cultures and habits of the workplace.

Ironically, all this attention to the "story" behind the report doesn't mean the report itself must be more complicated—in fact, it might make the report a simpler object since we can help students see it's only one piece of a larger puzzle (or, in Coe's terms, *ecosystem*).

Regularities of textual form, which are so easy to make into the focus of any assignment, are just an outcome of situated rhetorical purpose. That doesn't mean we shouldn't discuss formal characteristics of a genre; however, when we talk with students about those formal features, we should emphasize that formal qualities of the genre are *outcomes* of genre regularity rather than the defining characteristics of the genre. Whether or not I use genre language in a class, I always talk students through this relationship among a genre's situation, its purpose, and its formal elements. I respond to students' questions about how long a document should be, what it should look like, and other *what* questions by urging students to consider the *why*: Why might memos have the specific header information we generally see? Why might most recommendation reports begin by stating the conclusion? Why might direct address work well in most business letters while we find passive voice creeping in when the news is bad?

A rhetorical genre approach to professional communication (perhaps any kind of communication) complements teaching styles that prioritize exploration and questioning since there's actually something to talk about, something for students to discover. Genres shift our attention away from arcane textual patterns and rules and toward people working together in messy real-world situations to get things done. We don't end with description; the descriptive characteristics of genres are the starting point for classroom conversations about what those features might mean, how they might have come to exist, what their limits are, and how they serve particular purposes. Genre features are not necessarily logical in all cases, but they're never arbitrary—there's always a history and a context to unpack.

## CLASSROOM ACTIVITIES AND ASSIGNMENTS

There are two motives for teaching genre in the technical writing classroom: first, to help prepare students to write effective texts in particular genres, and second, to use the concepts and principles of genre to make sense of disciplinary and professional contexts.

*Genre Work in the Professions: Learning from Practicing Professionals*

Part of almost every introductory course in professional and technical communication, mine included, is an assignment asking students to learn about the writing habits of professionals in students' chosen fields, either by interviewing a professional or collecting examples of professionals' writing (or a combination of these activities). This assignment is a good chance to incorporate genre analysis using a simplified yet rhetorically based perspective on genre that takes into account the relationships between textual and contextual characteristics.

I ask students to identify and briefly describe their target workplace, identify someone they'd like to interview in that workplace, and then ask that professional for examples of the documents they write, either alone or (more often) with others. I advise students to ask their subjects about their routine or transactional writing (e-mails, notes, orders, meeting minutes), as well as more substantial or momentous writing since many professionals—even those who spend most of their time writing—don't think about their work as writing that's worthy of study.

Students then select one piece of writing they'd like to learn more about. I ask them to pick a genre that's repeatable (a genre for which there are multiple examples in this workplace) rather than unique (such as the company's website or its mission statement), which permits students to talk with their correspondent about aspects of the document that are *always there, sometimes there,* or *unique* to the present document.

Having students talk with a professional about a specific document written by that person (rather than just talking broadly about the person's working life) helps focus the conversation on actual practices and products; it almost always generates lots of how-we-do-things-here explanations. I give students some initial prompts but also encourage them to follow leads and ask questions about the document itself as well as about the document's relationship to the broader network of activities, people, and other documents that exist. Some of these questions include questions about exigence, audience and purpose, related documents, uniqueness and repetition, style, voice, tone, knowledge, and content. Here are a few of the starter questions I provide students:

- How did you know to write this document? Did it come from a previous document? Did someone ask you for it? Did you just realize for yourself that it was needed? (What led you to that assumption?)

- Who will read it? What will they do next? Where will the document go afterward (will it be filed, returned, passed along, discarded, revised)? Will other things be written in response?

- Is this document written in your style and your voice or is it more of a company-style or voice? Would someone be able to tell that *you* wrote it, or could anyone in this position have written it the same way?
- Why did you include [this piece of content]? Why did it go here? Where did it come from: your knowledge? research? some other source?

If the professional can provide other examples of similar documents, students might also ask questions about the similarities and differences among them.

Typically, I'll precede this exercise with a brief introduction to the principles of genre and some admonitions to look beyond the document's visible features and toward its functions, origins, and connections with other work. Although I don't expect students to master genre principles based on a short introduction, I want to shift their perspective toward the situational and rhetorical rather than formal, if only to improve their questioning when they talk with practicing professionals.

I should clarify that the purpose of the exercise is *not* to analyze a workplace genre, which would require a larger sample and greater immersion than we can achieve in a single assignment. Instead, the goal is to practice *thinking through genre*: to practice making sense of specific instances of genre work, spotting tacit practices and decisions that are part of professional activity, and making connections between an individual's efforts and the relationships, activities, and resources the individual's work is embedded in. Students come away from this activity not knowing terribly much about the genre they spent time on yet much more savvy about how a professional tackles writing tasks and how that activity fits into the professional's organizational context and work efforts.

### Comparing Multiple Examples of Familiar Genres

Regardless of how much we talk about the possibility of flexibility and choice among instances of a genre, it's hard to drive home that reality, particularly when our textbook offers a seemingly definitive and rationalized model. So, before I ask students to compose any texts in a genre, I attempt to demonstrate by example how much even very stable genres can vary while still getting the job done. The goal is to show how almost any aspect of a text, from big-picture variables (medium, form factor, organizational structure) to smaller details (voice, typography, argumentation, tone, manner of illustration), could be different without making the text an unsuccessful example of the genre.

I'm not looking for discipline-specific examples here but rather cases in which students recognize that the texts truly are examples of the same genre, despite their differences. I've used examples like travel guides (comprehensive guides, special-theme guides; online versus print guides; classic guides versus edgy or innovative examples; contemporary versus historical examples); grant proposals (ranging from short letter-style proposals to complicated, form-field-entry proposals for federal funds); textbooks; museum-exhibit programs; sales letters; shopping lists. I collect these things so I'll have them available to respond to questions about what a genre is "supposed" to look like to keep students from dwelling on the notion of the prototypical or ideal. I also ask students themselves to supply examples; any time we're working up to a genre-based assignment, we spend at least part of our time stepping back, considering the range and types of variations that might be possible and the quirks of context or environment that just might prompt a writer to use those variations.

This activity highlights not only how variable genres can be but—more important—also how thoroughly all rhetorical choices must be embedded in, and warranted by, the specific contexts and exigencies at hand. Texts vary, but they don't vary willy-nilly, and no choice is impact free. We look at oddball examples because they force us to imagine what situational elements or exigencies might have prompted that set of choices and what impacts might have been expected or achieved.

### Using Genre Sequences or Networks as Assignments

Whenever possible, I create assignments that relate in a sequence or cycle to demonstrate how most communication springs from other communication, leads to more communication, and draws upon or intersects with still other communication.

Of course, assignments are fundamentally "networked" in this way even if we make no special effort to highlight that fact: at the very least, students submit a piece of writing in response to an assignment sheet, which itself relates to a syllabus, and their submitted work will be "answered" by some form of evaluative communication. But I generally sequence assignments even more deliberately so that, as students make their way through my class, they'll see some of the same pieces of information, lines of reasoning, and even situational exigencies flowing through multiple examples of their own work, as well as the work of others (myself, at least, but also collaborators, peer reviewers, group members they might work with on a piece of research or for some other purpose).

## CONCLUSION

Much of what I use genre principles for in my teaching is pretty tacit and almost too simple to mention: I ask lots of questions about why a document looks or acts as it does; I ask questions about the students' major areas to get them to articulate what they know about them; I encourage them to introduce interesting examples of workplace or public writing so we can consider its generic qualities and relate them to what we know about its context. I try to encourage students, too, to shift their questioning (and their own answers to questions) toward genre matters since I hope this habit makes it more natural for them to view communication in these ways wherever they go. Likewise, by persistently viewing all documents as the "artifacts" of rhetorical contexts, artifacts that are intelligible to the people in those contexts, I attempt to do what Richard Coe (1994) urges instructors to do: to teach genre "as social process, archaeologically and ecologically" (163), as artifacts that help us to understand and act within the contexts where they're found.

My final recommendation is to be flexible but persistent in working with genre principles. Genre isn't so much a topic to teach as it is a way of conceptualizing writing as part of the larger system of resources and activities of a workplace. Genre is already inherent in students' work and our teaching; it needn't be added to your class as a new topic or unit. Instead, genre theory can be used to bring to the surface some of the otherwise hidden qualities of professional and technical writing and to help students connect their writing practices with the environments around them, now and in their future lives.

## DISCUSSION QUESTIONS

1. In what other contexts have you thought about genre, and what does genre mean in those contexts? How does that perspective on genre differ from the rhetorical genre studies perspective offered here?

2. Professional and technical communication courses often serve students from a variety of disciplines, including disciplines in the sciences, engineering, and business. All these disciplines feature writing conventions and expectations that likely differ from the conventions and expectations of your discipline. How might this chapter's perspective on genre affect your approach to working with students from less familiar disciplines?

3. Consider your own professional context—that of a teacher of professional or technical communication—in light of this chapter's discussion of genres. What genres are at work in this context (syllabi? institutional policies? assignments? grading rubrics? books such as the one you're

reading now?)? What would the genre set for this context look like, and how do those genres accomplish this community's purposes? Where do you fit into this system of activity?

4. If you have taught a common professional or technical genre before (memorandum, résumé, proposal, instruction set, brochure, etc.), how different or similar is the approach described in this chapter?

5. What was the last new genre you learned to write? How did you learn to write it? Does the genre perspective outlined here help you to think differently about that learning process? If so, in what ways?

## References

Artemeva, Natasha. 2006. "Approaches to Learning Genres: A Bibliographical Essay." In *Rhetorical Genre Studies and Beyond*, edited by Aviva Freedman and Natasha Artemeva, 9–99. Winnipeg: Inkshed.

Bakhtin, Mikhail. 1987. "The Problem of Speech Genres." In *Speech Genres and Other Late Essays*, edited by Michael Holquist and Caryl Emerson, translated by Vern W. McGee, 60–102. Austin: University of Texas Press.

Bawarshi, Anis S. 2003. *Genre and the Invention of the Writer: Reconsidering the Place of Invention in Composition*. Logan: Utah State University Press.

Bawarshi, Anis S., and Mary Jo Reiff. 2010. *Genre: An Introduction to History, Theory, Research, and Pedagogy*. West Lafayette, IN: Parlor.

Bazerman, Charles. 1994. "Systems of Genres and the Enactment of Social Intentions." In *Genre and the New Rhetoric*, edited by Aviva Freedman and Peter Medway, 79–101. Bristol, PA: Taylor and Francis.

Berkenkotter, Carol. 2001. "Genre Systems at Work: DSM-IV and Rhetorical Recontextualization in Psychotherapy Paperwork." *Written Communication* 18 (3): 326–49.

Berkenkotter, Carol, and Thomas N. Huckin. 1995. *Genre Knowledge in Disciplinary Communication: Cognition, Culture, Power*. Hillsdale, NJ: Lawrence Erlbaum.

Coe, Richard M. 1994. "Teaching Genre as Process." In *Learning and Teaching Genre*, edited by Aviva Freedman and Peter Medway, 157–69. Portsmouth, NH: Boynton/Cook.

Devitt, Amy J. 2004. *Writing Genres*. Carbondale: Southern Illinois University Press.

Devitt, Amy J. 2009. "Teaching Critical Genre Awareness." In *Genre in a Changing World*, edited by Charles Bazerman, Adair Bonini, and Débora Figueiredo, 337–51. Fort Collins, CO: WAC Clearinghouse.

Freedman, Aviva. 1993. "Show and Tell? The Role of Explicit Teaching in the Learning of New Genres." *Research in the Teaching of English* 27 (3): 222–51.

Freedman, Aviva, Christine Adam, and Graham Smart. 1994. "Wearing Suits to Class: Simulating Genres and Simulations as Genre." *Written Communication* 11 (2): 193–226.

Miller, Carolyn R. 1984. "Genre as Social Action." *Quarterly Journal of Speech* 70 (2): 151–67.

Orlikowski, Wanda, and JoAnne Yates. 1994. "Genre Repertoire: The Structuring of Communicative Practices in Organizations." *Administrative Science Quarterly* 39 (4): 541–74.

Russell, David R. 1997. "Rethinking Genre in School and Society: An Activity Theory Analysis." *Written Communication* 14 (4): 504–54.

Spinuzzi, Clay. 2003. *Tracing Genres through Organizations: A Sociocultural Approach to Information Design*. Cambridge: MIT Press.

Spinuzzi, Clay, and Mark Zachry. 2000. "Genre Ecologies: An Open-System Approach to Understanding and Constructing Documentation." *ACM Journal of Computer Documentation* 24 (3): 169–81.

Yates, JoAnne, and Wanda Orlikowski. 2002. "Genre Systems: Chronos and Kairos in Communicative Interaction." In *The Rhetoric and Ideology of Genre*, edited by Richard M. Coe, Lorelei Lingard, and Tatiana Teslenko, 103–21. Cresskill, NJ: Hampton.

# 6

# WHAT DO INSTRUCTORS NEED TO KNOW ABOUT TEACHING INFORMATION GRAPHICS?
## A Multiliteracies Approach

Karla Saari Kitalong

Jeff, a doctoral student at a small technological university, has been planning his first technical communication course for several weeks and feels quite confident about most of the major areas of coverage, including reports, correspondence, document design, project management, collaborative writing and editing, and technical style. With the support of his technical communication pedagogy professor and the other teaching assistants enrolled in the practicum, he has identified a community organization—a local food pantry—with which to collaborate, worked up a general framework for the course, and blocked out a series of carefully scaffolded assignments. The topic that remains to be planned is a unit on information graphics.

Jeff admits to procrastinating a bit on developing this unit. Although he understands the importance of visual displays of information in STEM fields and encountered both good and bad illustrations while reading psychology papers during his undergraduate studies, he doesn't have much experience with spreadsheets. Other than adapting an Excel grade book template for his own use in teaching, his spreadsheet experience stopped when he passed his high-school computer-literacy course, some ten years ago. He's concerned that his credibility as an instructor will suffer when his STEM students recognize they know more about spreadsheets than he does. He is also concerned that unless information graphics are carefully integrated into the work of the course, the students will treat them as busy work. Classes start right after the holidays. What can Jeff do to get up to speed?

DOI: 10.7330/9781607326809.c006

## THE STATE OF INFORMATION GRAPHICS IN
## TECHNICAL COMMUNICATION INSTRUCTION

Several years ago, I published a book chapter entitled "Select, Interpret, Produce: A Three-Part Model for Teaching Information Graphics" in which I argued that beyond teaching students to use words effectively, "technical communication teachers are expected also to teach information graphics," which require "design sensibility, software competence, and numerical fluency" (Kitalong 2007, 241). What more is there to say about information graphics now, nearly a decade after I contributed this pragmatic, pedagogically oriented chapter? Especially, what can be said that will help Jeff and other new teachers?

Information graphics remain important in technical communication and may even have increased in importance, both because readers are becoming more visually oriented (Brumberger and Northcut 2013) and because information is increasingly complex (Albers 2004; Mirel 2004). Thus, graphical representations that augment, or sometimes replace, detailed textual explanations are highly valued in the larger cultural context, as even the most cursory glance at news media outlets from *USA Today* to the *Wall Street Journal* will attest. Moreover, not only graphic designers and statisticians, but also professionals from virtually all walks of life, must, at one time or another, present graphical representations of their work. These representations must be clear so they convey as straightforward and coherent a message as possible. They should also be concise, free of what Edward Tufte called "chartjunk" (1983, 106), extraneous information that doesn't add to the meaning of the graphic. Finally, copyright and fair-use laws favor the use of information graphics produced by the author over the selection of information graphics found in published sources, unless such graphics are clearly labeled as copyright free or in the public domain.[1]

To date, plenty of technical communication teacher/scholars have addressed how to teach visual communication writ large (e.g. Bernhardt 1986, 1992, 1993; Bernhardt and Kramer 1996; Brumberger and Northcut 2013; Kimball and Hawkins 2008; Kostelnick 1995, 1996, 1998, 2008; Kostelnick and Hassett 2003; Kostelnick and Roberts 1998; Riley and Mackiewicz 2011). However, the specific topic of information graphics is a much less common pedagogical subject. Arguably, creating pie or bar charts, especially customized, rhetorically effective ones, requires not only design sensibility but also software competence and numerical fluency along with critical consciousness. Joanna Wolfe acknowledges a paradox inherent in our culture's interest in numerical relationships, linking them as much to rhetorical argument as to mathematical and statistical evidence: "On the one hand our culture tends to represent statistical evidence as a type of 'fact' and therefore immune to the arts of rhetoric, but on the other hand we are deeply aware and suspicious of the ability of statistics to be 'cooked,' 'massaged,' 'spun,' or otherwise manipulated. If statistics can be so altered by the method of their presentation—even as they continue to claim access to some sort of factual truth—aren't we clearly in the terrain of rhetoric?" (Wolfe 2010, 453).

In this chapter, I establish the importance of rhetorically constructing information graphics, as well as selecting, interpreting, and critiquing them, by situating information graphics within a larger context of increased visuality. I introduce Stuart Selber's multiliteracies model of technological literacy (Selber 2004), which is made up of instrumental, critical, and rhetorical literacies, as a framework for teaching information graphics in the technical communication classroom. Next, I describe the state of visual communication pedagogy as presented in three recent technical communication textbooks to demonstrate that the field as a whole has embraced a critical-literacy approach to information graphics. I conclude by suggesting a scalable framework for incorporating a multiliteracies pedagogy for information graphics within the established pedagogical framework to which we have become accustomed in technical communication.

## SELBER'S FRAMEWORK OF COMPUTER MULTILITERACIES

What skills and sensibilities does it take to develop, produce, and teach effective graphical representations? According to Selber, a "professionally responsible" approach in any context entails developing a pedagogy of "computer multiliteracies" from across a broad "conceptual landscape" (Selber 2004, 7, 25). Selber's landscape is shaped by three distinct but mutually constitutive literacies, which he calls "functional," "critical," and "rhetorical" literacy. Functional literacy is "organized by a tool metaphor" (24); usually, when computer users focus on acquiring functional literacy, their endgame is effective employment (25). Critical literacy considers computers to be cultural artifacts; critically literate people question assumptions related to technology, and their technology-related goal is objective critique. A rhetorically literate person views computers as "hypertextual media" (25); such people consider themselves to be not only users and critics of technology but also producers of "specialized literacy environments," including the relatively new type of highly interactive or metatextual environment that "connects other texts and their contexts in imaginative and meaningful ways" (136). The goal of the rhetorically literate communicator is reflective praxis (25).

Selber doesn't intend his framework to be hierarchical; that is, functional literacy is not a lesser literacy than is rhetorical or critical. The three literacies work together; thus, both teaching and learning in technological contexts—including the technical writing classroom—benefit from a situated pedagogy of multiliteracies that blends instrumental, critical, and rhetorical literacies. "Students who are not adequately exposed to all three literacy categories," he cautions, "will find it difficult to participate fully and meaningfully in technological activities" (Selber 2004, 24). Because "rhetorical literacy might be a particularly challenging place to start," he notes that it should not be valued more highly than functional or critical literacies. "There will be times when an attention to functional or critical concerns should be paramount" (25). Thus, information-graphics assignments in technical communication courses and programs must be carefully scaffolded so students have the opportunity to build their capacity in all three literacies. Neither teachers nor students, moreover, should aspire to full literacy in a given context; such a state is impossible to achieve, especially in a few weeks of a given semester.

For the typical technical communication teacher, critical literacy comes easily. We are practiced at analyzing and critiquing cultural phenomena, including technological products and processes, and at teaching students to deploy those critical tools in their work. Our field's rhetorical orientation gives us tools for determining what information is needed to make a persuasive argument aimed at a specific audience; that same rhetorical theory, applied to visuals, allows us readily to discern a well-designed and audience-oriented graphic. As the following section suggests, the textbooks we most commonly use to teach our courses also give prominence to critical-literacy approaches in the realm of information graphics.

## THE STATE OF INFORMATION-GRAPHICS PEDAGOGY IN TECHNICAL COMMUNICATION

Joanna Wolfe has argued for increased attention to quantitative argument in writing curricula, asserting that as "new technologies continue to increase the ease with which we can collect, compile, and compute large quantities of data, quantitative argument will come to play an even larger role in our daily lives" (Wolfe 2010, 453). Wolfe's call is worth considering, but for one reality: for the most part, technical communication faculty are trained in an English studies context. Thus, as a whole, we excel at "verbal literacy" but lack, "in approximately equal measure, design sensibility, software competence, and numerical fluency" (Kitalong 2007, 241). Our technical communication textbooks have not changed in ways that help us gain these missing sensibilities; in fact, a brief review of the information-visualization chapters of three current technical communication textbooks illustrates that their content has remained much the same as it was in 2007 when my earlier chapter was published. Thus, textbooks do not yet provide the exigency for technical communication teachers to develop novel pedagogical theories and strategies for teaching information graphics.

Although any three technical communication textbooks could have been examined for this exercise, Craig Rood, in his discussion of composition texts, suggests including "best-selling textbooks that have reached generations of students through multiple editions" (Rood 2013, 334); accordingly, I selected Mike Markel's *Technical Communication*, tenth edition (Markel 2012)[2] and Paul Anderson's *Technical Communication: A Reader-Centered Approach*, seventh edition (Anderson 2011). Rood further suggests that a "less-established textbook" might "offer a slightly different perspective" (Rood 2013, 334); Elizabeth Tebeaux and Sam Dragga's *The Essentials of Technical Communication*, third edition, fills the bill in this regard (Tebeaux and Dragga 2015).[3]

The three textbooks analyzed herein—and most other technical communication textbooks on the market today—present the entirety of visual communication in two—or at most three—chapters. Typically, one chapter addresses document design while a second emphasizes how to evaluate the design of graphics and illustrations and how to incorporate them into static technical documents. I will discuss the chapters on graphics and illustrations, leaving the chapters on page or document design for another time and place.

Mike Markel's textbook, *Technical Communication*, is typical of the top-selling textbooks in the field. That is, it covers practical topics in a critical-literacy vein, such as what to look for in selecting graphics for inclusion in technical documents and advice about the effective use of color, as well as more conceptual critical-literacy-related topics such as image ethics, especially in terms of citing images and avoiding reader deception. Markel also offers guidelines and checklists for addressing multicultural readers. Chapter 11, "Designing Documents and Web Sites," presents Markel's contribution to the theory and practice of document design, while chapter 12, "Creating Graphics," introduces graphics as the "'pictures' in technical communication," which include "drawings, maps, photographs, diagrams, charts, graphs, and tables" (Markel 2012, 306). Among Markel's unique contributions is a series of Tech Tips—such as "How to Insert and Modify Graphics" (313), "How to Use Tab Stops," "How to Create Tables" (324), and "How to Create Graphics in Excel" (325)—that offer instructions for using Microsoft's Office suite to create and manipulate graphics in technical documents.[4] This foray into functional literacy deploys an array of Tech Tips but does not include advice on spreadsheet use beyond the basic level. In fact, Markel's Tech Tips tend to emphasize design and exclude information on the skills or sensibilities needed for rhetorical production, such as the mathematical basis for graphics. His comment that graphics are "indispensable in demonstrating logical and numerical relationships" (307) tends toward critical rather than rhetorical literacy.

Another top-selling technical communication textbook, Paul Anderson's *Technical Communication: A Reader-Centered Approach*, eighth edition (Anderson 2014), is the only one of the three textbooks reviewed herein to offer three chapters on visual communication and the only one to place the two graphics chapters before the document-design chapter. Chapter 14, "Creating Reader-Centered Graphics," emphasizes the role of graphical elements in texts that rely primarily on the written word and sustains the book's theme of reader centeredness by providing graphics guidelines focused on attending first to how the writer can make the reader's task easier, more productive, and more fulfilling. Chapter 15, "Creating Eleven Types of Reader-Centered Graphics," was expanded in the eighth edition; previously it had been presented as a multipage checklist or set of guidelines sandwiched between the two visual-communication chapters. Chapter 15 is introduced with a table of typical tasks readers must accomplish, such as finding and using data, facts, or advice; understanding the relationship among variables; and comparing quantities. These tasks are matched with appropriate types of graphics (315). Anderson emphasizes reader centeredness throughout his textbook and foregrounds the primacy of writing perhaps even more than Markel does, but in general the two textbooks cover the same ground. Markel touches a little more on functional literacy than does Anderson, but both authors heavily emphasize the critical-literacy tactics of selection and interpretation, stopping well short of a full-fledged rhetorical-literacy perspective.

The third textbook, *The Essentials of Technical Communication* (Tebeaux and Dragga 2015), in its third edition, is neither as long lived nor as venerable as the other two. Its express goal, as articulated in the marketing-oriented front matter inserted into the teacher's edition, is to emphasize the "challenge of writing for readers who don't want to read" by providing a concise and cost-effective alternative to larger and more comprehensive texts. Affixed to the cover of the teacher's edition is a table that favorably compares its low price to the prices of the three top sellers, including Markel and Anderson. Unsurprisingly, however, the two visual-communication chapters—chapter 5, "Designing Documents," and chapter 6, "Designing Illustrations"—are virtually identical in content and scope to the chapters in the longstanding texts. For Tebeaux and Dragga, illustrations are classified into two types, tables and figures, with explanations of how to employ each most effectively. Tebeaux and Dragga's modest innovations include the introduction of video clips as sample illustration types; they also make a distinction between the primarily visual mode of illustrations and the multimodal nature of information graphics.

The graphical examples in all three texts are, of course, static, not only because they appear in print but also because current textbooks usually limit their coverage to static visuals: techniques for visually designing printed pages and informational websites, strategies for evaluating and selecting graphics for these types of presentations, and the ethics of representation. This range of topics—representing, once again, what Selber calls "critical literacy"—does not lend itself to understanding and manipulating numerical or statistical relationships or to developing and producing uniquely effective graphical representations.

## INFORMATION GRAPHICS IN TECHNICAL COMMUNICATION INSTRUCTION: TOWARD A PEDAGOGY OF MULTILITERACIES

We have seen thus far that teaching information graphics in the context of the technical communication course may seem daunting, especially to those of us who, like Jeff, have successfully avoided math for most of our careers. We tend to be comfortable with critique and therefore adept at enacting and teaching critical literacy, but functional and rhetorical literacy may seem outside our grasp. Meanwhile, the textbooks that support our course designs include limited scaffolding for teaching information graphics, even as our audiences' expectations and learning styles demand a visual approach.

In *Designing Texts*, Eva Brumberger and Kathryn Northcut incorporate all three components of Selber's conceptual landscape of computer literacy by emphasizing visual literacy, which is arguably almost as foreign as numerical literacy to many English-studies-trained faculty. They define visual literacy as consisting of "looking, seeing, thinking, and producing" (Brumberger and Northcut 2013, 3). "Focus on student-produced texts rather than quizzes, tests, or theory," they urge their readers, because what is important is the "students' ability to understand and apply concepts" (6). They recognize, too, that teaching visual communication is a "lot of hard work" but also "fun" and perhaps "addictive" (6). They advocate that teachers embrace the challenge. I argue in this chapter (although Brumberger and Northcut don't stress this) that for technical communication teachers, embracing the challenge entails developing a level of functional, critical, and rhetorical literacy sufficient to internalize the unique combination of design sensibility, software competence, and numerical fluency required for designing and teaching effective information graphics in the context of a technical communication classroom.

Selber's pedagogy of multiliteracies has the advantage of being scalable. That is, a teacher can design a single class around multiliteracies, or a faculty can design an entire program with such a framework in mind. At the institutional level, a multiliteracies approach could render a commitment to a student learning outcome of quantitative literacy much more robust.[5] The key in any of these contexts is to provide the necessary structure or scaffolding. The following recommendations provide a starting point for Jeff and other new teachers of technical communication—a way to prepare for this work. Of course, preparation takes some time, depending upon your initial experience and time available.

### Develop or Update Your Personal Functional Literacy

Teachers are constantly updating their software skills and knowledge. The most common spreadsheet program, Excel, is exceptionally robust, easy to learn, and widely available. Training programs, too, are readily available: your university or college may offer workshops or short courses, although these are often directed at office workers and thus may not be suitable for educators.[6] If you prefer to learn from a book, you will find dozens of current offerings. Or, you can easily find online tutorial sites. For example, skilledup.com serves as a portal for a wide range of tutorials categorized by subject area. In the data science category, an article entitled "6 Excel Gurus That Will Help You Become an Expert" offers a selection of free online courses.[7] I "test-drove" the chandoo.org site, subtitled "Become Awesome in Excel," which included materials that seemed to be at the right level for me, but of course it might take some time and experimentation to find a good fit among the thousands of available online resources.

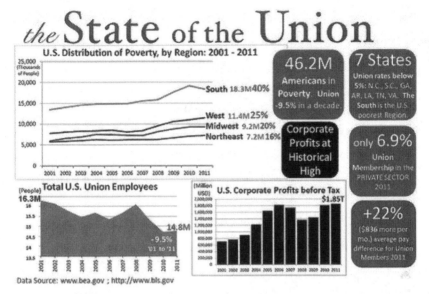

*Figure 6.1. This graphic describes poverty distribution by region as it relates to income levels of union employees and corporate profits. (Graphic by Cory M. Grenier https://www. flickr.com/photos/26087974@N05/. Used under a Creative Commons attribution license. Downloaded November 14, 2016, from https://www.flickr.com/photos/26087974@N05/ with/7987644934/.)*

*Syllabus and Assignments*

Carefully go over your syllabus and assignments to identify points at which instruction about information graphics might be introduced. For example, data could be construed as a source of primary research, even though textbook authors might not acknowledge this. Certainly, the credibility of an argument is enhanced by providing supporting or opposing evidence from primary as well as secondary and tertiary sources; however, Markel's catalog of types of primary research includes "observations, demonstrations, inspections, experiments, interviews, questionnaires, or other field research" (Markel 2012, 123). Might closely reviewing and manipulating statistics compiled by relevant organizations or governmental agencies also be considered a type of primary research? Might information graphics like the one shown in figure 6.1, located by means of a Creative Commons search, be considered as a research source?

Once relevant data sources are found, students can identify and download appropriate data sets. The data can then be manipulated in Excel to create information graphics that provide visual support for their particular arguments. This activity combines the functional literacy

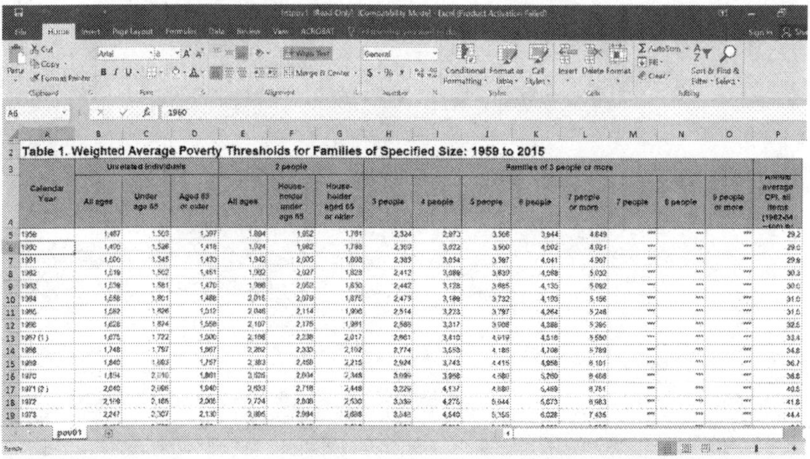

Figure 6.2. Sample dataset showing poverty levels as related to family size and makeup, 1959–2015. (Downloaded from www.census.gov. http://www.census.gov/data/tables/time-series/demo/income-poverty/historical-poverty-people.html.)

needed to download and manipulate data in Excel with the rhetorical literacy of presenting visual arguments precisely and persuasively.

General demographic data that is in the public domain can be found at the US Census Bureau's rich site (http://www.census.gov). Recall that Jeff's students will be developing promotional, procedural, and instructional materials for a local food pantry. To understand part of the context of food pantries, Jeff could assign his students to research the history of family poverty in the United States by downloading the Excel table shown in figure 6.2. This table compiles historical data on the relationship between family size and poverty levels.

Armed with tabular data spanning more than fifty years, students could create charts that illustrate patterns of poverty over time for families with specific demographic characteristics; figure 6.2 shows the table with all data prior to 2000 deleted and specific columns highlighted. Figure 6.3, then, shows a very rough line graph that tracks changes in poverty levels for two-person families with heads of household aged sixty-five and older. With data tables such as these, Jeff could begin to experiment with creating illustrative and persuasive charts and graphs; obviously the one shown in figure 6.3 needs editing!

Depending on the theme of the class, other data sources can be found. For example the US Bureau of Labor Statistics documents employment trends in a variety of technical fields including engineering and computer science. A teacher might use the "National Occupational

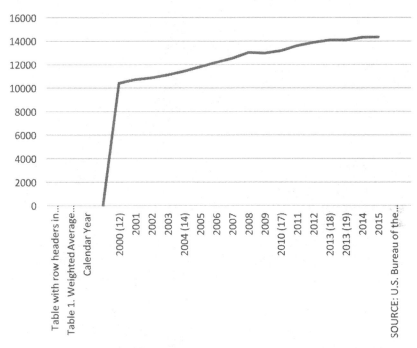

*Figure 6.3. The first pass at an information graphic Jeff created from data provided by the US Census Bureau. Clearly, the graph needs editing; Jeff will spend some time producing successive drafts that better tell a story of escalating poverty among the elderly, and he will incorporate an attitude of experimentation into the assignments for his class. (Source for data: www.census.gov. Graph designed by the author.)*

Employment and Wage Estimates" data, complete with a downloadable Excel file at http://www.bls.gov/oes/home.htm. Still another class might engage students in projects associated with their local college or university context. All public colleges and universities compile statistics about students, faculty, and staff that can be retrieved and imported into a spreadsheet. For example, at my institution, enrollment statistics can be requested from a staff member and pasted into an Excel spreadsheet.[8]

By reviewing basic Excel tools and locating relevant data sources that complement the theme of the class, Jeff completes the groundwork needed to incorporate a multiliteracies approach to information graphics into his course. As he thinks about the skills, critical sensibilities, and rhetorical principles that will help his students, he comes to the conclusion that a unit on information graphics is best integrated with other lessons, just as one might integrate the legal, ethical, and intercultural content found in most of today's technical communication textbooks.

## DESIGNING ASSIGNMENTS TO STRENGTHEN FUNCTIONAL, CRITICAL, AND RHETORICAL LITERACIES

Jeff resolves to start with the familiar: early in the semester, he asks students to collect examples of poverty-related information graphics from news media or online sources. They critique their found graphics using classificatory information presented in their textbook and then evaluate them using the textbook's design guidelines, which often emphasize technical rather than rhetorical features. As they apply their critical-literacy skills, they may recognize that some information graphics they find to be effective deviate from the textbook author's guidelines; similarly, technically correct information graphics may not be persuasive, and graphics can skew data, as any technical communication textbook is sure to point out. This critical-literacy activity provides an excellent opportunity for discussing all three literacies—functional, critical, and rhetorical—as they relate to the perceived effectiveness of information graphics.

When students create their own graphics from raw data, ethical issues may be brought out; thus, ethics should not be ignored in discussions about information graphics. You need not go beyond the textbook you are using for ways to accomplish this. For example, designing and producing data displays that compare different numerical relationships from the same table can illuminate alternative interpretations of data that has been presented as straightforward and unproblematic. And one of Markel's Tech Tips on designing effective bar graphs with Excel, though seemingly focused on functional literacy, includes ethical and rhetorical suggestions as well: "Make the proportions fair. Make your vertical axis about 25 percent shorter than your horizontal axis. An excessively long horizontal axis minimizes the differences. Make all the bars the same width, and make the space between them about half as wide as a bar" (Markel 2012, 326). The details enumerated in this guideline—including the appropriate amount of space between bars and the ratio of horizontal to vertical axis—are quite technical; executing and manipulating graphics at this level of detail requires skill and practice with the software. But the guidelines also help enact critical literacy; for instance, they can be used as criteria for evaluating existing bar charts. To support the development of students' rhetorical literacy, such guidelines can be followed (or, for that matter, resisted) in students' own infographic designs.

If you think in terms of *infusing* your course with a multiliteracies approach to information graphics instead of limiting yourself to a more conventional stand-alone unit on information graphics, your own and your students' confidence and competence may be strengthened. In short, you may be able to teach information graphics more thoroughly when you consider them as embedded within the larger array of communication theories and practices important to your technical communication course.

### FOR FUTURE CONSIDERATION

If the coverage of visuals in the field's textbooks is any indication, technical communication's interest is limited to static information graphics. However, the world at large needs the kind of information visualization that, in the words of TED Talk speaker David McCandless, enables us to "see the patterns and connections that matter" and encourages us to develop designs that allow information to "make . . . more sense, . . . tell . . . a story," or encourage us to "focus only on the information that's important" (McCandless 2010).

For McCandless and other observers of the contemporary communication scene, information graphics are increasingly interactive. An interactive information graphic, writes Wibke Weber, is a "visual representation of information that integrates different modes," including "image (which is the constitutive element), written text, sound, layout," to form a "coherent whole." In addition, "at least one navigation option" allows users/readers "to control the graphic." An interactive information graphic has a distinct "communicative function": instead of an often static visual representation of information or data, an interactive information graphic informs by, for example, "describing or explaining something or narrating a factual story" (Weber 2015).

Although interactivity can be had simply by asking the reader to click on a thumbnail image to enlarge a graphic or launch a video,[9] at even the most basic level of interactivity, "every click should provide access to new information or to the next information sequence" (Weber 2015).[10] At higher, more elaborate levels of interactivity, "users can influence or modify the content and choose their own navigation path through the information graphic" (Weber 2015). It is this latter type of interactivity to which students in technical communication classes have become accustomed by means of their smartphones and tablets; despite their familiarity, however, they may be discouraged from including such graphics in their technical communication course reports, not to mention that they may lack the software skills to produce such elaborate graphics. A multiliteracies approach to teaching information graphics—one that encourages students' concurrent and integrated development of functional, critical, and rhetorical literacies—is the best preparation, not only for working with static graphics, as expected in current technical communication pedagogical practice, but also for considering and producing fully interactive representations of complex data relationships.

## Notes

1. See, for example, http://libguides.mit.edu/usingimages. Also visit the Creative Commons site (https://creativecommons.org) for a wide range of sample graphics that can be used and modified.
2. The eleventh edition was published in 2014.
3. Tebeaux and Dragga are as well known as Anderson and Markel in the field of technical communication; in fact, Sam Dragga has written extensively on ethics in visual presentation, both in collaboration with Dan Voss (Dragga and Voss 2001, 2003) and on his own (Dragga 1992). Moreover, Tebeaux and Dragga have each authored several textbooks, both individually and together. The first edition of *The Essentials of Technical Communication* was published in 2010, several years after the pedagogy book containing my aforementioned chapter.
4. My purpose is to describe rather than to critique the three selected textbooks, but I do question the usefulness of Tech Tips, given that unless they are written in rather generic terms, they would quickly become dated; conceivably, the technical advice could be obsolete before the book is printed. Functional literacy related to the production of information graphics is necessary, but software tutorials seem superfluous in a printed textbook and might dictate more frequent textbook updates at increased cost to students. A friendlier solution might be to provide Tech Tips in an easily updated format such as a supplemental website.
5. As a starting point, see the Quantitative Literacy rubric developed by the Association of American Colleges and Universities (AAC&U) at http://www.aacu .org/sites/default/files/files/VALUE/QuantitativeLiteracy.pdf.
6. Online tutorials aimed at professors are often geared to creating Excel productivity tools such as grade books and attendance trackers. Microsoft's Educator Community (https://education.microsoft.com/) is one such site.
7. See https://digital.com/blog/excel-tutorials/.
8. See, for example, http://www.admin.mtu.edu/em/services/erlstat/index.php?qtr =fall2014andreport=bandmap=falseandsubmit=1.
9. See, for example, the *New York Times* information graphic *The Danger of Digging Deeper*, which is a video that describes earthquake dangers associated with drilling miles into the earth to reach sources of geothermal energy (http://www.nytimes .com/interactive/2009/06/23/us/Geothermal.html).
10. A good example is the interactive periodic table of the elements offered by the Los Alamos National Laboratory (http://periodic.lanl.gov/index.shtml), in which hyperlinks take readers to an additional, very detailed screen or two of information about the element.

## References

Albers, Michael J. 2004. *Communication of Complex Information: User Goals and Information Needs for Dynamic Web Information.* Mahwah, NJ: Lawrence Erlbaum.

Anderson, Paul. 2011. *Technical Communication: A Reader-Centered Approach.* 7th ed. Boston, MA: Cengage Learning.

Anderson, Paul. 2014. *Technical Communication: A Reader-Centered Approach.* 8th ed. Boston, MA: Cengage Learning.

Bernhardt, Steve. 1986. "Seeing the Text." *College Composition and Communication* 37 (1): 66–78.

Bernhardt, Steve. 1992. "Revisions: Seeing the Text, Again." *Journal of Computer Documentation* 16 (3): 48–53.

Bernhardt, Steve. 1993. "The Shape of Text to Come: The Texture of Print on Screen." *College Composition and Communication* 44 (2): 151–75.

Bernhardt, Steve, and Robert Kramer. 1996. "Teaching Text Design." *Technical Communication Quarterly* 5 (1): 35–60.

Brumberger, Eva, and Kathryn M. Northcut. 2013. *Designing Texts: Teaching Visual Communication.* Amityville, NY: Baywood.

Dragga, Sam. 1992. "Evaluating Pictorial Illustrations." *Technical Communication Quarterly* 1 (1): 47–62.

Dragga, Sam, and Dan Voss. 2001. "Cruel Pies: The Inhumanity of Technical Illustrations." *Technical Communication* 48 (3): 265–74.

Dragga, Sam, and Dan Voss. 2003. "Hiding Humanity: Verbal and Visual Ethics in Accident Reports." *Technical Communication* 50 (1): 61–82.

Kimball, Miles A., and Ann R. Hawkins. 2008. *Document Design: A Guide for Technical Communicators.* Boston, MA: Bedford/St. Martin's.

Kitalong, Karla Saari. 2007. "Select, Interpret, Produce: A Three-Part Model for Teaching Information Graphics." In *Resources in Technical Communication: Outcomes and Approaches,* edited by Cynthia L. Selfe, 241–64. Amityville, NY: Baywood.

Kostelnick, Charles. 1995. "Cultural Adaptation and Information Design: Two Contrasting Views." *IEEE Transactions on Professional Communication* 38 (4): 182–96.

Kostelnick, Charles. 1996. "Supra-Textual Design: The Visual Rhetoric of Whole Documents." *Technical Communication Quarterly* 5 (1): 9–33.

Kostelnick, Charles. 1998. "Conflicting Standards for Data Displays: Following, Flouting, and Reconciling Them." *Technical Communication* 45 (4): 473–82.

Kostelnick, Charles. 2008. "The Visual Rhetoric of Data Displays: The Conundrum of Clarity." *IEEE Transactions on Professional Communication* 51 (1): 116–30.

Kostelnick, Charles, and Michael Hassett. 2003. *Shaping Information: The Rhetoric of Visual Conventions.* Carbondale: Southern Illinois University Press.

Kostelnick, Charles, and David D. Roberts. 1998. *Designing Visual Language: Strategies for Professional Communicators.* Needham Heights, MA: Allyn and Bacon.

Markel, Mike. 2012. *Technical Communication.* 10th ed. Boston, MA: Bedford/St. Martin's.

McCandless, David. 2010. "The Beauty of Data Visualization." Filmed July 2010 at TEDGlobal. TED video, 18:10.

Mirel, Barbara. 2004. *Interaction Design for Complex Problem Solving: Developing Useful and Usable Software.* San Francisco, CA: Morgan Kaufman.

Riley, Kathryn, and Jo Mackiewicz. 2011. *Visual Composing: Document Design for Print and Digital Media.* Upper Saddle River, NJ: Prentice Hall.

Rood, Craig. 2013. "Rhetorics of Civility: Theory, Pedagogy, and Practice in Speaking and Writing Textbooks." *Rhetoric Review* 32 (3): 331–48.

Selber, Stuart A. 2004. *Multiliteracies for a Digital Age.* Carbondale: Southern Illinois University Press.

Selfe, Cynthia L., ed. 2007. *Resources in Technical Communication: Outcomes and Approaches.* Amityville, NY: Baywood.

Tebeaux, Elizabeth, and Sam Dragga. 2015. *The Essentials of Technical Communication.* 3rd ed. New York: Oxford.

Tufte, Edward R. 1983. *The Visual Display of Quantitative Information.* Cheshire, CT: Graphics.

Weber, Wibke. 2015. "What Is an Interactive Infographic?" Conference material prepared for Malofiej 23, March 15–20, 2015. http://www.malofiejgraphics.com/what-is -an-interactive-information-graphic/.

Wolfe, Joanna. 2010. "Rhetorical Numbers: A Case for Quantitative Writing in the Composition Classroom." *College Composition and Communication* 61 (3): 452–75.

# 7

# DESIGNING TEACHING TO TEACH DESIGN

Eva Brumberger

You've just been given a copy of the updated outcomes for the technical communication program and asked to ensure that your course is designed to address the new student learning outcomes. As you skim the document, you see that *visual communication* appears in several places, with phrases like, *communicate verbally and visually in multiple genres, understand and apply visual design principles,* and *design documents for specific audiences and purposes.* The semester starts in a few weeks, and you've already put together the syllabus for your course, complete with assignments and a tentative schedule. You've included a unit on visual communication, but as you review your course objectives and assignments, you realize design really isn't getting much attention at all. With a deep sigh, you reconcile yourself to redoing the syllabus. But what will you need to add in order to address the visual communication learning outcomes? And the course is already packed—how in the world are you going to fit more in without taking out other material that is also important?

Designing information has become central to the work of the technical communicator; it is at the heart of websites, online help and reference documents, presentations, promotional materials, and, yes, now even paper-based manuals and reports (see Brumberger 2007 for a brief history of this growth). Many more programs require a course in document/information design than did previously—40 percent in 2011 as compared to 4 percent in 2005 (Meloncon and Henschel 2013), which is an enormous step forward. Further incorporating visual communication into existing technical communication courses would extend this significantly and would help students recognize that design is integral to all facets of their work.

But, our courses are already overflowing with content, and for many of us, design remains unfamiliar—and uncomfortable—terrain, which makes it even more difficult to know what to add, let alone how to add it.

DOI: 10.7330/9781607326809.c007

The goal of this chapter is to connect that terrain to the more familiar landscape of teaching writing and to provide conceptual and pragmatic tools that will help instructors incorporate information design into existing course content. Ultimately, the goal is to help you design your teaching so visual communication becomes an integral part of all your technical communication courses—so you actually teach design while you are teaching other central technical communication concepts. The chapter will introduce important concepts behind the teaching of design, address some basic principles of document design, offer strategies for integrating design concepts and principles into general technical communication courses, and suggest assignments (and assignment "makeovers") that can help students become better visual thinkers and problem solvers. So, overall, the chapter answers two important questions:

- What should I be teaching students about visual communication within the context of my other technical communication courses?
- How can I incorporate this conceptual and practical material into my courses?

## IMPORTANT CONCEPTS

Teaching visual communication must begin with helping students conceptualize design as rhetorical rather than decorative (see, for example, Brumberger 2004; Buchanan 1995; Hill and Helmers 2004; Joost and Scheuermann 2007; Kostelnick and Hassett 2003). That is, like verbal language, visual language shapes users' interactions with an information product (a print document, a website, a blog, etc.): it can affect perceptions of ethos, set tone, impact clarity, and so on, ultimately playing an important role in whether a product works for its intended users and achieves its anticipated purpose.

Teaching design also involves dispelling the long-standing notion (see Brumberger 2007) that design is "esoteric, mystical magic" (Dondis 1973, 184)—the product of some sort of idiosyncratic, black-box affair that is largely unteachable because it relies on intuition to the exclusion of analytical reasoning, or what Rudolph Arnheim (1980) calls "intellectual cognition." Talent and intuition help, of course, just as they do with writing. However, again like writing, design is an analytical problem-solving process; to be successful, it must rely on both intuitive and intellectual cognition (Arnheim 1980). Notice that there are two familiar terms here: *problem solving* and *process*. Both are key for teaching design.

Design, and particularly the type of design we do in technical communication, is about solving a communication problem (see Carnegie

[2013] for a detailed discussion of design as problem solving). Making important information easier to find, making a complex task clearer and less intimidating, making the navigation of a website more apparent—all these are problem-solving tasks that depend upon design. Again, design is not about making an information product (a document, website, blog, etc.) more aesthetically appealing, although that is often a secondary goal; it's about making that product more effective for its intended audience and purpose. Of course, as Donald Norman (2004) has argued, aesthetics and effectiveness are often intertwined. We are more inclined to use products that appeal to us, which means they are that much closer to achieving their purpose. Still, no matter how attractive a document is, if audiences can't find the information they need, it ultimately fails in its purpose.

How does design problem solving occur? It starts with looking at—and learning to see—a problem in new ways. We rely on patterns as we navigate our lives, and we form habits based on those patterns; these habits can be incredibly useful because they can enable us to get things done more efficiently and with less mental energy. At the same time, though, they can work against creative thinking and problem solving. Here's a very basic example: suppose a company has always used left-justified headings to distinguish among sections in the hard-copy reports they produce for investors. The company wants to keep hard-copy reports but wants them to be more attention grabbing and more user friendly. Habit might suggest simply changing fonts, colors, sizes, and so on for the headings. More creative problem solving, on the other hand, might suggest doing all that but also rotating the headings 90 degrees and running them down the outside edge of the page; this approach would allow users to quickly locate sections by fanning the edge of the report, and it would suggest a more creative and forward-thinking ethos for the company.

Analysis of the design problem is followed by planning, idea generation (brainstorming, sketching, storyboarding), drafting, and revision—stages that once again parallel the writing process and similarly may not occur in a linear progression. Foregrounding these parallels helps students understand that design is not entirely alien. It certainly asks them to rely on a different mode of thinking—visual instead of verbal—but the same is true of subjects like math and science. And, as several scholars have argued, success as a technical communicator now depends in large part on the flexibility to use different modes of thought and communication (Brumberger 2007; Hocks and Kendrick 2003; Johnson-Sheehan 2002; Stroupe 2000).

## BASIC DESIGN PRINCIPLES

The concepts discussed previously shape the way design is taught; they are the underlying foundation for visual communication pedagogy. Layered on top of that foundation are the principles at work in design itself, the principles students need to learn in order to craft effective visual communication.

Many principles that guide design are based on Gestalt theories of visual perception. Steven Bradley (2014) of *Smashing Magazine* provides an accessible summary of the central tenet of Gestalt theory: "We see the whole as more than the sum of the parts, and even when the parts are entirely separate entities, we'll look to group them as some whole." The principles linked to this tenet describe the ways we organize visual information in order to form the whole. There are any number of resources that provide extensive discussion of these principles. *The Non-Designer's Design Book* (Williams 2014) offers a quick, fun, and affordable introduction to design principles that is both accessible and engaging. For more depth, but still a practical approach and friendly, relaxed style, take a look at *White Space Is Not Your Enemy* (Hagen and Golombisky 2013). And, of course, a simple Google search returns scores of helpful sites and visual examples galore. Here, I provide a quick overview of principles that are derived from Gestalt theory and figure prominently in the types of design work technical communicators do.

### Grouping

One of the key ways in which we organize visual information is through grouping. Several perceptual principles support grouping, but two of the most important are proximity and similarity. Proximity simply means we tend to perceive things that are close together as being parts of a group; the nearer they are to one another, the more closely they are grouped. As can be seen in figure 7.1, proximity implies a relationship.

The principle of similarity leads viewers to group items together regardless of their proximity. Each item on a page (print or digital) has numerous characteristics: size, shape, color, position, texture, and so on. Any of these characteristics by itself may become the basis for grouping by similarity. So, for example, even when items on a page are separated by distance, aligning them connects them visually, thereby providing organization. Of course, the characteristics can also work together to make grouping even stronger. In the left image in figure 7.2, for instance, the eye tends to group the As together, even though they are

*Figure 7.1. Grouping through proximity; we see the people in figure 7.1b as having a closer relationship than those in figure 7.1a because of their proximity to one another.*

```
A  B  B  B  B  B  B  B  B  B          A  B  B  B  B  B  B  B  B  B
B  A  B  B  B  B  B  B  B  B          B  A  B  B  B  B  B  B  B  B
B  B  A  B  B  B  B  B  B  B          B  B  A  B  B  B  B  B  B  B
B  B  B  A  B  B  B  B  B  B          B  B  B  A  B  B  B  B  B  B
B  B  B  B  A  B  B  B  B  B          B  B  B  B  A  B  B  B  B  B
B  B  B  B  B  A  B  B  B  B          B  B  B  B  B  A  B  B  B  B
B  B  B  B  B  B  A  B  B  B          B  B  B  B  B  B  A  B  B  B
B  B  B  B  B  B  B  A  B  B          B  B  B  B  B  B  B  A  B  B
B  B  B  B  B  B  B  B  A  B          B  B  B  B  B  B  B  B  A  B
B  B  B  B  B  B  B  B  B  A          B  B  B  B  B  B  B  B  B  A
```

*Figure 7.2. The principle of similarity leads us to group the As even though they are distributed across the array; when the font is changed, the grouping becomes stronger.*

distributed across the array and don't stand out in any way other than letter shape. As the image on the right in figure 7.2 shows, when we change the font, thereby providing additional visual cues, the grouping becomes more pronounced.

### Contrast

Similarity is not the only principle at work in figure 7.2; the grouping in the image on the right is also stronger because of the principle of contrast, which serves to separate dissimilar items. In the right-hand image, the font used for the As has serifs, and the letterforms are wider. These distinctive factors help the As stand out from the Bs. Adding boldface, or changing the size or color of the As, would make the contrast still stronger, thereby enabling us to group information more quickly and easily. Contrast is essential for distinguishing different types of items from one another, and it is a tool for providing visual emphasis; it is

Figure 7.3. On the left, gray text on a gray gradient makes for weak figure-ground contrast. On the right, strong contrast makes the text easier to read.

also what enables us to perceive an item as distinct from its background (figure-ground contrast), as shown in figure 7.3.

### Balance

Like grouping and contrast, balance brings visual order to the page. The balance of a page depends on the distribution of positive space—the parts of the page that include "objects" (text, images, and so on)—and negative space (also referred to as *white space*)—the unoccupied areas of the page. The way in which elements are placed on the page also determines what receives emphasis, what users perceive to be most important, and even how they navigate the page.

Notice that design does not need to be symmetrical to achieve balance (see fig. 7.4). In fact, an asymmetrical but balanced design is often more effective because it is more dynamic and is more interesting to the eye.

### Repetition

Repetition refers to the consistent and repeated use of design elements within a design—typefaces, typeface attributes (bold, italic, etc.), colors, alignments, spacing, and so forth—which creates visual patterns and rhythms that help users navigate a document. These consistent patterns unify a design and contribute to visual clarity, while subtle differences across elements can be confusing or jarring to the eye. For example, if

*Figure 7.4. The star and three circles of increasing size at the right counterbalance the large circle in the upper left of this asymmetrical design.*

one random letter B in figure 7.2 had a slightly different alignment, the viewer might notice it and ask, "Why is that one different?" That distraction can interfere with the rhetorical goals of the design. Similarly, if some of the main-level headings in this chapter were slightly smaller than others, or were followed by slightly less space, readers might think they were looking at a different type of heading that was less important. Repetition thus also enables the designer to create a clear visual hierarchy on the page, which helps users access and comprehend information more quickly and easily.

### The Whole Is More Than the Sum of the Parts

Grouping, contrast, balance, and repetition are at the heart of effective design, but like the principles of effective writing, they don't function in isolation. First, the principles are interconnected, so application of one principle affects how the other principles are working. For example, changing the way two elements contrast with one another can also change the balance of the page. Second, design principles work together not only to ensure visual clarity and conciseness but also to determine visual style, set tone, and build ethos. Go back to contrast for a moment. Combining a lot of bold contrasts makes for a dynamic, high-energy page. This combination might be great for convincing kids to sign up for soccer. However, it wouldn't be so effective for helping Medicare recipients choose a health plan, a complex and information-laden process in which the ethos of the medical provider is a key component.

In short, the whole is more than perceptual; designers must consider both the perceptual impact and the rhetorical impact of their design choices. Last, but most definitely not least, what users perceive when they look at the page depends not only on perceptual and rhetorical principles but also on their own prior experience and expectations.

## INTEGRATING DESIGN INTO COURSES

As I noted at the beginning of the chapter, figuring out *what* to teach students about visual communication is only half the battle; the other half is figuring out *how* to teach it within existing contexts and constraints. If you look at typical technical communication textbooks, they include discrete chapters on visual communication. The temptation is to follow this structure and teach design as a unit. Don't do it. Teaching design as a unit relegates it to an add-on—to delivery in the most stripped-down, rhetorically impoverished sense. It sends a message to students that design comes after writing instead of sending the message that design begins in the thinking and planning stages of a document.

Visual communication must be integrated throughout the course, much as attention to audience pervades each topic and assignment. This approach helps students recognize that verbal and visual communication work together. What does this mean in practical pedagogical terms? It means recrafting syllabi to weave in visual communication from the learning objectives onward. Learning objectives for a general technical communication course might include a statement as straightforward as this one, which could easily be adapted for other courses: *Understand and apply basic principles of visual communication in order to design documents that are effective for their intended users.* In order to meet this objective, design must be introduced early in the course as foundational material; I try to bring design into classroom conversation within the first three weeks of a fifteen-week course. Once that happens, it is visible and expected to varying degrees in all the activities and assignments that follow.

## INTEGRATING DESIGN INTO ACTIVITIES AND ASSIGNMENTS

Incorporating visual communication into existing assignments instead of adding design-specific projects has three pedagogical benefits: first, it reinforces the centrality of design to technical communication; second, it means students have more opportunities throughout the course to apply the design principles they learn; and third, it encourages students to consider the ways in which verbal and visual rhetoric can (and should) work together to make documents more effective. And there's one additional benefit: it means you don't have to add more projects to your already full courses in order to address visual communication learning objectives and outcomes.

Any number of the activities and assignments that often appear in technical communication courses can easily be adapted to include visual communication. My goal here is not to provide a comprehensive list of assignments, accompanied by details about how you might teach each one. Instead, I'd like to walk you through some examples that illustrate pedagogical patterns you can tailor to particular assignments in your own courses.

Take, for instance, a basic rhetorical analysis, for which students are asked to examine a document and discuss the ways in which rhetorical strategies are used. This is an assignment that might come early in the course, before students begin crafting their own documents. Along with its discussion of verbal rhetoric, the analysis could include an examination of visual rhetoric: How does the design contribute to the ethos? How is color used to make the document more persuasive? What messages do the images seem to convey? In what ways does the visual style seem matched to the intended audience and purpose? How do the visual style and the verbal style support one another? And so on. I should emphasize that, although I've separated out design-centered questions here in order to provide ideas, the visual aspects of the analysis should be interwoven with the verbal aspects rather than addressed on their own. So, in its exploration of ethos, the analysis would discuss both verbal *and* visual factors; likewise, any discussion of style would address both verbal *and* visual style, as well as the ways they interact. The structure of the analysis, then, encourages students to make connections between verbal and visual rhetoric rather than treating them as distinct and unrelated.

As discussed elsewhere (Brumberger 2005), job-application materials, too, are excellent projects in which to practice design skills. The effectiveness of a résumé, especially, is extremely dependent upon its visual communication. Yet all too often, students rely on Microsoft Word templates for their résumés, or they adhere to samples from campus career services, which tend to be generic at best. We spend time in class analyzing the strengths and weakness of each, and then I ban both these options. Students must then think about and articulate factors that should determine the design of their résumé, starting with the quantity and type of information they need to include and the sort of job for which they are applying (e.g., a user-experience specialist could get away with a much less traditional résumé than could an accountant). They also must reflect on what they want the design to convey about them (e.g., should it suggest that they are creative? conservative?); this reflection opens a conversation about how they might apply different design elements and principles to achieve their goals. For example, the two résumés in figure 7.5 contain exactly the same information. Figure 7.5a is generic in its design. Nothing stands out as particularly important, the reader cannot extract information by scanning quickly, and it is difficult to sort out how things are grouped in the experience section. Figure 7.5b has exactly the same content but makes use of basic design principles to provide a stronger visual organization. Because of stronger contrast, more purposeful use of grouping, and the addition of features such as bullets, a potential employer can immediately see important information, such as the skills and past job titles that may set the candidate apart from other applicants. The résumé is still extremely conservative in its use of contrast (e.g., only one font is used throughout), but it is has a stronger professional ethos and better showcases the applicant's strengths.

A résumé chock full of marketable skills and wonderful job experience will not get the applicant an interview if potential employers must hunt for the information. Similarly, if employers have two applicants with comparable qualifications, the applicant whose résumé conveys a professional and polished ethos is more likely to get the interview. Including a class exercise in which you walk through an analysis of the effectiveness (and ineffectiveness) of several different résumés can help students begin to recognize that the power of a résumé is created in large part through its visual design.

Larger technical communication projects likewise offer multiple opportunities for analytical and hands-on design work to accompany writing. Think about an instruction-writing project paired with usability testing, for example. When students analyze the needs of their intended audience for the instructions, they should also articulate how users might read the instructions and in what context; these factors should shape the design, not just the writing. For instance, if users are going to be on the shoulder of an interstate trying to change a tire for the first time, they're not going to want to turn pages in the instructions or have them blow away in the wind. Something like a stand-alone laminated quick-reference card would be much more helpful than having to hunt through the owner's manual. And, the users will probably be stressed

Figure 7.5. A generic résumé (fig. 7.5a) made more effective with the application of basic design principles (fig. 7.5b).

out and impatient, so being able to glance at the instructions and quickly find the necessary steps is essential. As with the résumé, the use of contrast and grouping will be particularly important. In this scenario, though, the design is not about getting readers' attention and conveying a professional ethos; it's about helping them complete a task as quickly and safely as possible. There are many opportunities to introduce these ideas early in an instructions project. For example, you can ask students to bring in instructions they think are (in)effective, and have the class to articulate the ways in which visual design contributes to or detracts from the effectiveness. You can also provide students with examples and ask them to redesign the examples so the visual communication better addresses the intended audience and context of use.

As students work on their instructional documents and move from analysis to planning to drafting, I continually remind them that they must design the instructions (visually and verbally) to best suit the needs and contexts of their audience. And, of course, when students test their instructions, they again must pay attention to the interplay between design and usability. Crafting user personas can enrich the project further. If feasible, an even more effective approach is to recruit actual users to participate in the design-and-development process, which can be enabled by careful selection of project topics.

Similarly, formal-report projects lend themselves well to teaching design. As I've noted previously (Brumberger 2005), many types of professional reports rely substantially on visual communication, both in their use of images that support the verbal content and in their overall design. Formal reports typically include front matter (e.g., a table of contents), recurring items such as headers and footers, multiple levels of headings, and so on. No matter how small each design decision may seem, it contributes to the overall visual rhetoric of the report. One way to emphasize this is again by beginning the project with analysis: share models of professional reports and ask students to analyze the rhetorical role—and effectiveness—of the design, as well as the verbal text. This analysis can lead into students' analysis of the goals for their own reports and, in turn, to more careful consideration of design choices during the planning and drafting stages.

Thus far, I've discussed assignments that focus on more traditional genres of technical communication. Digital-information products, of course, provide another valuable set of design opportunities, and these can complement and enrich what students learn from designing print documents. Websites and related digital products lack the production constraints of print documents (e.g., limited use of color due to printing costs). At the same time, though, they introduce visual issues that are not concerns for print design, such as type rendering, dynamic image content, fluid page sizes, and flexible navigation, to name just a few. The same foundational design principles still hold, but their application becomes ever more complex as the number of design elements and technological considerations multiplies. That said, students who can move comfortably between print and web—as writers and designers—will ultimately be much more attractive to employers, so projects that ask this flexibility of them are particularly valuable.

The assignments discussed here represent only a small subset of the types of work that happen in technical communication courses. However, they demonstrate that design can be integrated simply and seamlessly into existing projects if you introduce it early in a course and approach it rhetorically.

## Evaluation

Of course, once you've integrated visual communication into class projects, you must also assess students' learning. Evaluating design work can be particularly uncomfortable because of a lingering sense that it is somehow more subjective than evaluating writing. The discomfort is, not

| | Excellent | Good | Needs Improvement | Unacceptable |
|---|---|---|---|---|
| Applies design principles to indicate relationships & hierarchies among items | | | | |
| Enables users to navigate the document quickly & easily | | | | |
| Uses contrast to set items apart & emphasize material as appropriate | | | | |
| Uses repetition and consistency to establish visual patterns | | | | |
| Controls negative space to balance the page & guide users | | | | |
| Uses color effectively & consistently | | | | |
| Uses fonts effectively for both readability & aesthetic impact | | | | |
| Has an overall look/feel that is unified and conveys an appropriate ethos | | | | |

*Figure 7.6. General introductory rubric for evaluating visual communication.*

surprisingly, even more pronounced when you are new to visual communication. However, as Kathryn Northcut (2013) reminds us, "Choosing the most appropriate criteria for evaluation must begin with asking what we want students to learn" (190). When we return to the idea that design fulfills a rhetorical purpose rather than being merely decorative, we can develop effective evaluative criteria (see DeVoss [2013] for more discussion of evaluation and assessment of visual communication), which can be incorporated into assignment rubrics.

A general introductory rubric for evaluating the design of technical communication might look something like figure 7.6. Notice the rubric includes some rhetorical factors but is heavily weighted toward perceptual principles. This is a fine starting place, but as students begin to master basic design principles, the rubric should include more rhetorical criteria.

As the semester progresses and projects get more complex, the rubric should also include more criteria that are matched to the assignment. For example, a résumé relies heavily on typography but often does not use color; the rubric, then, might include more detail about typography use and might omit color entirely.

Asking students to reflect upon their work is also very helpful in teaching design, not necessarily for assigning a grade, per se (although it can certainly be used to support grading efforts), but because it can offer additional insights into students' thought processes and conceptual understanding that may not be immediately apparent from examining the project drafts and final copy themselves. As important, guided self-reflection can help students grow as designers, just as it can help them grow as writers.

### Tools and Technologies

There is one last aspect of teaching design that bears mention here, and that is the role of technology. As Eva Brumberger, Claire Lauer, and Kathryn Northcut maintain, "Visual communication requires students (and practitioners) to balance an understanding of complex conceptual principles with the use of equally complex software" (Brumberger, Lauer, and Northcut 2013, 173). Design today is heavily reliant on sophisticated and powerful tools, particularly layout software (e.g., ADOBE INDESIGN and FRAMEMAKER), graphics software (e.g. ADOBE ILLUSTRATOR), and image-editing software (e.g., ADOBE PHOTOSHOP). Should technical communication students learn to work with these tools? Well, yes. However, even within the context of a dedicated visual communication course, it is challenging to balance teaching concepts and principles with teaching the tools to support them. Within other technical communication courses, it becomes even more difficult because visual communication is only one of many strands of learning. So, I would not advocate for dedicating a lot of time teaching tools specific to visual communication. I would, however, argue strongly for introducing students to foundational tool concepts and equipping them with a basic level of technology skills they can build on in future courses.

At the very least, students should be introduced to the use of styles, whether through MICROSOFT WORD, ADOBE INDESIGN, or some other word-processing or layout software. Styles support consistent visual design within a document, but they also encourage students to think about the structure and organization of information more generally, which is valuable whether or not the pedagogical focus is visual communication. Additionally, if students learn to define styles (for headings, bulleted lists, body text, etc.) and tag text (apply the styles) consistently for print-based documents, they will be well-positioned to grasp concepts and skills required in more advanced technical communication courses, such as writing for the web.

Teaching the use of styles in MICROSOFT WORD, can serve as an entry to the more powerful capabilities of layout software, such as ADOBE INDESIGN. If time allows, you might ask students to take a document they have created in MICROSOFT WORD, and redesign it with layout software. Students will quickly recognize the complexity of using the more sophisticated software, but they will also see how many more options it opens up for design. There are many online tutorials (some free, others through subscription) that can help students extend their skills with design software without using large chunks of classroom time.

Ultimately, there is no one approach to technology that fits all teaching contexts; instead, you must determine what will work within the framework of your course, your access (and that of your students) to various tools, your student population, and your programmatic context (Sheppard 2013).

## CONCLUSION

Although teaching visual communication may seem a daunting undertaking at first, in many ways it parallels the more familiar work of teaching writing. In this chapter, I've tried to highlight those parallels, beginning with the critical concept that design is a process of rhetorical problem solving. I've also provided a brief overview of foundational design principles. My goal here is to help those who are new to design begin to develop a level of comfort and knowledge that will allow them to teach introductory design with confidence. As I've noted, there are a lot of user-friendly resources for additional learning and support. And finally, with a focus on typical assignments in general technical communication courses, I've offered suggestions for integrating visual communication into existing course content, including approaches to evaluating that work. While this chapter certainly cannot (and should not!) be considered a comprehensive guide to teaching design in the general technical communication classroom, I hope it will help you develop your own pedagogical goals and strategies as you start thinking about design in new ways. Design is a discipline in which the act of looking is essential—looking at design all around you in daily life, as well as models from experts in the field. The more you (and your students) look, the more you will learn to see design principles at work and, in turn, the more you will be able to teach and apply those principles.

## DISCUSSION QUESTIONS

1. Collect examples of several documents you think are well designed or poorly designed. What caused your initial reaction/judgment? Now, look for the design principles at work in each document. Can you explain your initial reaction in terms of the concepts and principles discussed in the chapter?

2.  How do you see the perspectives discussed in this chapter shaping your own pedagogical decisions in terms of ideology, goals, assignments, and evaluation?

3.  What assignments do you envision including in the course(s) you will be teaching? Using the ideas from the chapter as a starting place, how might you make design a part of those assignments?

4.  Think about what you want students to learn about design in your course(s). How will you evaluate their learning? What strategies or approaches that you use to evaluate writing might be applied to evaluating design?

5.  Based on your own experiences—as a student, a teacher, and/or a practitioner—how would you approach teaching technology? What contextual realities might stand in the way of your approach?

## References

Arnheim, Rudolph. 1980. "A Plea for Visual Thinking." *Critical Inquiry* 63 (3): 489–97.

Bradley, Steven. 2014. "Design Principles: Visual Perception and the Principles of Gestalt." *Smashing Magazine*, March 29, 2014. https://www.smashingmagazine.com /2014/03/design-principles-visual-perception-and-the-principles-of-gestalt/.

Brumberger, Eva. 2004. "The Rhetoric of Typography: Effects on Reading Time, Reading Comprehension, and Perceptions of Ethos." *Technical Communication* 51 (1): 13–24.

Brumberger, Eva. 2005. "Visual Rhetoric in the Curriculum: Pedagogy for a Multimodal Workplace." *Business Communication Quarterly* 68 (3): 318–33.

Brumberger, Eva. 2007. "Making the Strange Familiar: A Pedagogical Exploration of Visual Thinking." *Journal of Business and Technical Communication* 21 (4): 376–401.

Brumberger, Eva, Claire Lauer, and Kathryn Northcut. 2013. "Technological Literacy in the Visual Communication Classroom: Reconciling Principles and Practice for the 'Whole' Communicator." *Programmatic Perspectives* 5 (2): 3–28.

Buchanan, Richard. 1995. "Rhetoric, Humanism, and Design." In *Discovering Design: Explorations in Design Studies*, edited by Richard Buchanan and Victor Margolin, 23–68. Chicago, IL: University of Chicago Press.

Carnegie, Teena. 2013. "Design as Problem Solving." In *Designing Texts: Teaching Visual Communication*, edited by Eva Brumberger and Kathryn Northcut, 33–48. Amityville, NY: Baywood.

DeVoss, Danielle. 2013. "Evaluating and Assessing Designed Documents: Assignments, Projects, Portfolios, and More." In *Designing Texts: Teaching Visual Communication*, edited by Eva Brumberger and Kathryn Northcut, 219–40. Amityville, NY: Baywood.

Dondis, Donis. 1973. *A Primer of Visual Literacy*. Cambridge: MIT Press.

Hagen, Rebecca, and Kim Golombisky. 2013. *White Space Is Not Your Enemy: A Beginner's Guide to Communicating Visually Through Graphic, Web & Multimedia Design*. 2nd ed. Burlington, MA: Focal.

Hill, Charles, and Marguerite Helmers. 2004. *Defining Visual Rhetorics*. Mahwah, NJ: Lawrence Erlbaum.

Hocks, Mary, and Michelle Kendrick. 2003. "Introduction: Eloquent Images." In *Eloquent Images: Word and Image in the Age of New Media*, edited by Mary Hocks and Michelle Kendrick, 1–16. Cambridge: MIT Press.

Johnson-Sheehan, Richard. 2002. "Being Visual, Visual Beings." In *Working with Words and Images: New Steps in an Old Dance,* edited by Nancy Allen, 75–96. Westport, CT: Ablex.

Joost, Gesche, and Arne Scheuermann. 2007. "Design as Rhetoric—Basic Principles for Design Research." Paper presented at the 3rd Swiss Design Network Symposium, November 17–18, 2006, Geneva, Switzerland.

Kostelnick, Charles, and Michael Hassett. 2003. *Shaping Information: The Rhetoric of Visual Conventions.* Carbondale: Southern Illinois University Press.

Meloncon, Lisa, and Sally Henschel. 2013. "Current State of U.S. Undergraduate Degree Programs in Technical and Professional Communication." *Technical Communication* 60 (1): 45–64.

Norman, Donald. 2004. *Emotional Design: Why We Love or Hate Everyday Things.* New York: Basic Books.

Northcut, Kathryn. 2013. "Evaluating Visual Communication." In *Designing Texts: Teaching Visual Communication,* edited by Eva Brumberger and Kathryn Northcut, 181–96. Amityville, NY: Baywood.

Sheppard, Jennifer. 2013. "Balancing Act: A Guide to Analyzing Context and Developing a Technologically Appropriate Approach to Visual Communication Instruction." In *Designing Texts: Teaching Visual Communication,* ed. Eva Brumberger and Kathryn Northcut, 245–64. Amityville: Baywood Publishing Company.

Stroupe, Craig. 2000. "Visualizing English: Recognizing the Hybrid Literacy of Visual and Verbal Authorship on the Web." *College English* 62 (5): 607–32.

Williams, Robin. 2014. *The Non-Designer's Design Book.* San Francisco, CA: Pearson Education/Peachpit.

# 8

# DESIGNING AND WRITING PROCEDURES

David K. Farkas

Professor Sakina Hussain is beginning the unit on procedures ("how-to" discourse) in her introductory course in technical communication. She knows procedure writing is an essential part of such a course. Because procedure writing is one of the most straightforward genres of professional communication, she likes to cover procedures early in the course.

Sakina first asks students to list how many kinds of procedures (in any medium) they encounter in a typical week or two. (Often she and the students use the term *instructions* as well as *procedures*. *Instructions*, however, does not have a useful singular form and therefore is at times awkward to use.) Student responses include printed product instructions, online help systems for software products, short video tutorials posted on eHow.com (and elsewhere), recipes, face-to-face instructions of all kinds (e.g., When the road forks, turn right), and much more. Sakina has her own examples ready to display so if a student calls out "computer manual" or "video," she can give the entire class a quick look at one of the examples and make a few introductory comments. She wants students to recognize right away that amid great diversity, the typical procedure consists of a title, an introductory paragraph or two, and a list of numbered steps.

To sharpen students' concept of procedures, Sakina points out closely related forms of discourse. Process descriptions are similar to procedures but explain something, such as photosynthesis, that happens without a human agent. Product demos (often videos) also resemble procedures—and may even show how to use the product—but their main purpose is to explain the product's features and benefits, not how to use it. Functional descriptions are a borderline case. If the phrase "Pressure release" is stenciled next to the hand wheel of a gate valve in an industrial setting, it is assumed the operator knows what action to take to relieve the pressure, so there is no directive about turning the

DOI: 10.7330/9781607326809.c008

wheel. Similarly, when a computer user mouses over a toolbar button, the tooltip may read "Increase font size," but there is nothing about selecting text and clicking the button.

Sakina tells students that professionals in many fields create procedures or are responsible for products and services that will not be successful if their accompanying procedures are not usable. An understanding of procedures, therefore, is valuable to a wide range of professionals, as well as to technical communication professionals.

Sakina encourages students to voice their complaints about frustrating procedures they encounter because these complaints will get them thinking about good and bad design and help them realize that while procedures appear to be easy to write, considerable craft is required to write effective procedures.

## KEY CONCEPTS FOR UNDERSTANDING AND WRITING PROCEDURES

Now, moving into the heart of her unit on procedures, Sakina deepens students' understanding of procedures by explaining some key concepts:

1. **Domain of a procedure**: the kind of system being explained

2. **System state**: the focus on the assembly, operation, maintenance, or repair of a system

3. **Audience (users)**: especially the audience's particular background and information needs but also general characteristics widely shared among human beings

4. **Medium and modalities**: in particular, whether the procedure employs text, text with graphics, or video

These concepts prepare students to make effective design decisions as they encounter the diverse circumstances in which procedures must be written. For example, they learn that the domain of natural systems often presents special challenges for writing steps, that malfunctioning systems require special troubleshooting procedures, that audiences may or may not want procedures designed to promote retention, that the narration in video procedures focuses more on concepts than on steps, and that video presents drawbacks as well as benefits. For insightful analyses of digital products based on a conceptual framework with categories similar to these, see S. Scott Graham and Brandon Whalen (Graham and Whalen 2008).

Later, Sakina will explain procedures as a set of mandatory and optional components, an approach that breaks down the task of writing a procedure into a series of smaller, more manageable writing decisions. Finally, she will assign exercises and out-of-class assignments built around her approach to teaching procedures. Here, then, are the concepts Sakina presents.

## The Domain

Much procedural discourse is written to explain how to use digital systems (such as software applications, websites, or smartphones). Although procedures for digital systems may be conceptually difficult ("Configuring Glype on your proxy server"), digital systems employ a carefully crafted graphical user interface (GUI) that greatly simplifies the task of writing the steps. Even if a manual or help system contains a hundred or more procedures, the user repeatedly performs similar operations (clicking, tapping, dragging, scrolling, etc.) on the same few types of interface elements: tabs, menus, dialog boxes, option buttons, and so forth.

Mechanical and electrical equipment requires the user to interact with physical components such as switches, latches, and fittings (although such systems may also include a GUI display). Various complications arise, such as explaining how tight to fasten something and when a part is worn out.

Explaining how to carry out tasks in the natural world is often difficult. Consider, for example, instructions for shucking an oyster. An oyster is not designed to be opened up with a knife. There is no user interface. It is quite challenging—even with video—to make clear how to grasp the oyster, how to find the part of the shell that covers the hinge, where to insert the knife point, and how to work the knife with the correct twisting motion. Similarly, it is challenging to create instructions for serving a tennis ball or hitting a golf ball off a tee. For tasks such as these, visuals, and ideally videos, are almost a necessity. See Robert Krull (2001) on procedures focused on the natural world.

There are also procedures for intangible domains, such as how to apply for a transfer (and other administrative tasks) and how to remember the names of people you meet.

## The System State

Most systems exist in more than one state. Therefore, we write procedures for getting a system going (assembly, installation, configuration), for operating the system (routine and emergency operation), for maintaining the system, and for repair following a malfunction. Often when systems malfunction, the problem is not obvious, so trouble-shooting procedures with special diagnostic components are necessary. Troubleshooting procedures first guide users in identifying the problem and then provide one or more solutions to that problem. For an introduction to troubleshooting procedures focused on digital systems, see David K. Farkas (2010).

## The Audience

Understanding your audience and adapting to their background and information needs is central to writing procedures, as it is to all forms of communication. The background of the intended audience, in particular their level of experience and expertise, in large part governs their information needs—what terms and concepts must be explained, the pacing of the procedure, and more. For more advanced audiences, the pacing can be faster, and you can write higher-level (and therefore fewer) steps than are necessary for entry-level (or novice) audiences. A difficult problem is how to write procedures for audiences with very different backgrounds. For a useful conceptual treatment of the backgrounds, abilities, and information needs of audiences, see Michael J. Albers (2003).

One important dimension of an audience's background is culture. Instructions for products sold worldwide require metric (SI) as well as English units of measurement. Also, if a manual or video shows people using a product, you should take care that their clothing, behavior, and other elements of the visual do not offend or confuse people from certain cultures. For a broad theoretical examination of localization, see Miguel A. Jimenez-Crespo (2013).

An important dimension of information needs is whether to design for retention. Procedures in first-aid manuals are optimized for fast reading and immediate action. But for training EMTs (emergency medical technicians), who must provide first aid without consulting procedures, special *training* procedures are prepared. These training procedures include previews, reviews, quizzes, and other elements that promote retention. Note also that first-aid procedures are routinely classified as members of the genre first-aid manual. However, because first-aid procedures usually begin with diagnosis (is there breathing? are the lips blue?), they can be classified conceptually as repair/troubleshooting procedures for a natural system, the human body.

When we think about audience, we usually think about differences. But we can also consider characteristics that are essentially universal, based on hard-wired aspects of human perception and information processing. All human beings have limitations to their short-term memories, and these limitations fall within a well-established range, with few outliers. Therefore, in all our design and writing decisions, we must be aware of the cognitive load we are creating, and we must strive to reduce this cognitive load. See Mariana Coe (1996) for an explanation of human perception and information processing as applied to technical communication.

Another general characteristic of users is that they benefit from and typically appreciate well-designed examples. Because examples are concrete and (as much as possible) draw upon the users' previous experience, they can reduce cognitive load. You can embed a brief example in a single numbered step. Tutorial documentation employs one or more lengthy examples, sometimes called *scenarios*, as the narrative framework into which multiple procedures, along with components for retention, are embedded.

Furthermore, human beings have competing demands on their time and are often impatient when reading procedures, especially product instructions. As discussed by John M. Carroll and Mary Beth Rosson, users are tempted to bypass instructions even though this decision is often counterproductive (they would ultimately have done better reading the instructions) (Carroll and Rosson 1987). Students readily acknowledge this impulse of impatience. Because of this human characteristic, we must strive to make our procedures fast and easy to work with, make them visually appealing, and perhaps find ways to make them interesting and enjoyable. Conversely, procedures with a confusing visual design and poor writing not only increase the user's cognitive load but immediately lose credibility and make users more apt to set them aside.

### The Medium and Modalities

A communication medium is the means through which content is communicated through space and time. The newspaper is one of the print media. Modalities are symbol systems through which information is encoded: voice, text, visuals, video, and sign language are all modalities. Particular modalities are possible or not possible in particular media. Text, for example, is possible in print, on the web, and even in cinema (for example, the scrolling text used for credits) but is not possible in traditional telephony. Digital media can do more than just convey modalities. They permit many forms of interactivity, such as the display and hiding of blocks of content, hyperlinking, and search.

*Text and Visuals*

Most procedures communicate primarily through the modality of text. Often it is highly beneficial to supplement text with visuals. Visuals can be purely conceptual (for example, a flowchart). Often, however, visuals show the appearance of a device (or the parts of the device) or its user interface, and they indicate where and possibly how to carry out an action. Visuals can also show what the device or system should look like when the action has been carried out successfully (feedback).

There are many considerations in the design of visuals. Photographs are easy to create and show realistically what an object looks like. However, drawings, in particular line drawings, provide a more filtered view: the artist can show only what is important for the purpose of the visual. Consider, for example, the potential problem of unwanted shadows in a photograph. Sakina knows that tech-savvy students may challenge this distinction, making the point that photographs can now be converted to line drawings with varying degrees of fidelity to the original photograph.

Another issue is the view. Are we looking at the device from the front, top, or side? Do we want an exploded view? How much of the device should be included? A broad view often makes it harder for the viewer to locate the element of interest within the visual. But if the view is too narrow (just a close-up), the viewer loses context and may not be able to locate the element represented in the visual on the actual device. Still another consideration is the coordination of the visual with the text, including the effective use of captions and callouts.

To minimize translation costs, manufacturers sometimes prepare procedures that consist entirely or almost entirely of visuals. This approach may work for the assembly of a simple bookshelf, but it rarely works for more complex products. (See Richard Mayer [2005] for a broad survey of the uses of text and visuals [both static and motion visuals] in learning.)

Sakina often assigns students to find procedures and bring them to class for group discussion. The groups analyze the strengths and weaknesses of the procedures, including the visuals. Then, each group reports their most interesting observations to the class as a whole. Class discussion of badly written procedures can be instructive and fun. However, be restrained in making fun of bad writing that arises simply from the writer's deficiencies in English.

*Video*

The web and other digital media allow the use of narrated videos— which we divide into (1) photorealistic videos (recorded with a video camera or smartphone), (2) drawn animations, and (3) recordings of activity on a computer screen. If we look back to the examples of shucking an oyster and swinging a golf club, we see that photorealistic videos and drawn animations are beneficial for documenting tasks that (1) take place in the natural world and (2) involve motion. Drawn animations have advantages over photorealistic videos similar to the advantages of drawn graphics to photographs.

Narrated screen recordings are often used to document digital systems. Adobe, for example, has created libraries of video procedures to teach Photoshop, Dreamweaver, and other products. When such video procedures are well designed, users benefit from receiving information through both their visual and auditory channels. Because the visual channel is very good at showing the location of interface elements, how to use them, and what the result is (feedback), the narration can focus on the purposes of these actions, which is hard to communicate visually. Even so, we should not assume video procedures are clearly superior to print procedures, especially because users do not have the complete control of videos (and other time-based media) that they have with text and static graphics (Alexander 2013). Rereading a sentence is fast and easy; replaying a segment of a video is not.

Sakina sometimes assigns students to make quick videos, using their smartphones, of tasks performed in the natural world. She also tells students they can easily make their own narrated screen recordings with freeware such as Screencast-O-Matic.

### Voice Only

Finally, Sakina takes note of voice-only instructions. In both our nonwork and professional lives, we are often called upon to give oral instructions (although we can often grab a pencil or even "draw in the air" with our fingers). Furthermore, we now have automobiles that can conduct an interactive dialogue with the driver.

### TEACHING HOW TO WRITE PROCEDURES

After introducing the key concepts that underlie procedures, Sakina explains in detail how to write procedures. First, she wants her students to understand the function of the basic procedure components: the title, introductory paragraph, steps (including feedback statements), and notes (see fig. 8.1). Second, she wants them to understand the function of the four kinds of steps (Farkas 1999; van der Meij and Gellevij

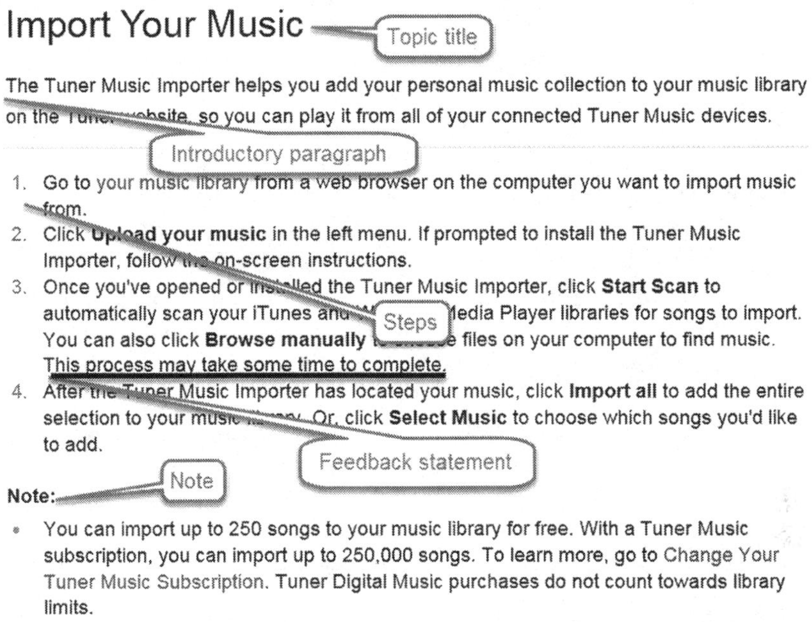

# Import Your Music — Topic title

The Tuner Music Importer helps you add your personal music collection to your music library on the Tuner website, so you can play it from all of your connected Tuner Music devices.

Introductory paragraph

1. Go to your music library from a web browser on the computer you want to import music from.
2. Click **Upload your music** in the left menu. If prompted to install the Tuner Music Importer, follow the on-screen instructions.
3. Once you've opened or installed the Tuner Music Importer, click **Start Scan** to automatically scan your iTunes and Media Player libraries for songs to import. You can also click **Browse manually** — files on your computer to find music. This process may take some time to complete.
4. After the Tuner Music Importer has located your music, click **Import all** to add the entire selection to your music — Or, click **Select Music** to choose which songs you'd like to add.

Steps

Feedback statement

**Note:** — Note

- You can import up to 250 songs to your music library for free. With a Tuner Music subscription, you can import up to 250,000 songs. To learn more, go to Change Your Tuner Music Subscription. Tuner Digital Music purchases do not count towards library limits.

*Figure 8.1. A typical procedure showing the most widely used components.*

2004). Unsurprisingly, the construction of procedures varies greatly, and students will, for example, encounter steps that are combinations and hybrids of the kinds of steps described here. However, the approach I describe not only enables students to write effective mainstream procedures but also helps both students and instructors understand the construction and function of any procedure they encounter.

## The Title and Introductory Paragraph

The title identifies the procedure and briefly explains its purpose. When a simple consumer product such as a bookshelf ships with a single assembly procedure, the title can be very general—perhaps *Instructions* or *Assembly* or *How to Assemble Your Ikea X-15 Bookshelf.* But if the procedure is part of a large set of procedures, such as the many help topics that make up an online help system, the titles must be more specific and informative so users can choose the procedure they want when numerous titles appear together in a table of contents or some kind of digital picklist.

The title is often followed by an introductory paragraph (or paragraphs). Often this paragraph includes more detailed information about the purpose of the procedure than the title can provide. This enables the user to confirm that this procedure does indeed meet their goals. The paragraph may also state some secondary consequence or side effect the user may not want. For example, if a user saves a document created in a word-processing application as a .txt (ASCII) file, the file size is much reduced, and there are other benefits as well. However, because some users may not know (or may not call to mind) that all formatting will be lost, the introductory paragraph of a procedure entitled "Saving your work as a .txt file" should warn the user about losing formatting. (The application should also display a system message with a similar warning.) For a mechanical task, the introductory paragraph might include such useful information as the time required and the tools and parts that will be needed. Writing a truly useful introductory paragraph requires a broad understanding of user needs.

### Steps and Feedback

Following the title and the introductory paragraph (if present), most procedures transition to a list of numbered steps, written as commands. The primary role of the steps is to enable the user to execute the procedure they have by now decided on. We can therefore say the title and introduction are about decision making and steps are mainly about execution.

Procedures can also be written in paragraph form with descriptive (indicative) verbs. A sixteenth-century treatise on mining and smelting (Agricola 1556) is written this way: "The tiny flakes . . . of silver adhering to stones or marble or rocks . . . should be smelted in the furnace of which the tap-hole is only closed for a short time." Descriptive procedures typically place an extra cognitive load on users, so they are much less prevalent than procedures built around imperative verbs. At times, however, a writer seeks to embed extensive descriptive information within a procedure, and then a descriptive procedure becomes a good choice.

Sakina explains that there are four different kinds of steps, each with a different function. She classifies steps as (1) action steps (the simplest kind of step), (2) user-option steps, (3) conditional steps, and (4) local-purpose steps. There are also feedback statements that directly follow a step.

An action step directs the user to carry out a particular action: "Press the **On** button." But action steps may also include the location of the interface element to be acted upon. In software documentation, many steps begin with such location information ("On the **Home** tab, in the **Styles** group, click . . ."). It may also be necessary to make clear *how* to carry out the step, especially in the domain of mechanical devices: "Release the catch mechanism by pushing firmly on the ends of both latches." An action step may include information on what not to do or warn against a bad way of carrying out the step: "Do not tighten the bolt excessively."

Almost any kind of step can be supplemented by a feedback statement. Feedback statements describe the result of carrying out an action: "Tap any note in your document. Blue boxes appear around all the notes." Feedback statements assure the user that they have done the right thing and that the system is responding properly. The feedback statement in the figure (above) assures the reader that the system is responding properly despite the slow response time the user may experience. Because feedback statements make procedures longer, they should be used sparingly.

User-option steps are distinctive in that they throw the user back into decision-making mode by inviting the user to switch to a variation on the main goal of the procedure. For example, "If you prefer a crispy-crust pizza, do not use a baking sheet and bake the pizza at 375 degrees." It would be entirely possible—but very inefficient—to write two separate procedures, one for regular-crust and one for crispy-crust pizzas. A user option does not have to be formatted as a separate step. Instead, it can be included with another step. The option for making a crispy-crust pizza might well have been included with the step for making a regular crust.

Sometimes the writer knows users may encounter an impediment (an unwanted condition) when trying to carry out a procedure. Conditional steps explain how to identify this unwanted condition and then address it: "If you are in the Transactions view, switch to the Report view." Note that if the user is indeed in the Transactions view of this software product, they will not be able to carry out any procedure pertaining to reports unless they address the condition by making the switch. Feedback statements are often conditionals: "If red triangles appear in the window, adjust the positioning knob until black triangles appear." The need for numerous conditional steps in a set of procedures suggests either that the system is prone to malfunction or that its user interface is poorly designed.

The last kind of step we should teach students in an introductory technical communication course is the local-purpose step. In most cases, a local-purpose step is built upon another kind of step but adds an extra phrase (usually beginning with *to*) that explains why the step is necessary: "Rap periodically on the filter housing to ensure maximum suction." By explaining the purpose of an action, we increase the likelihood that the user will carry it out. Other local-purpose steps serve as milestones in a long procedure: they sum up for the user how much they have completed: "Click **Bottom** to complete the first box."

### Cautions, Notes, and Tips

Cautions, notes, and tips call special attention to an item of information. Cautions and warnings warn about actions that might harm human beings, damage systems, or result in unexpected loss of data. Many cautions and warnings are conditionals: "**Warning**: If the temperature indicator reaches the red zone, increase coolant flow immediately."

Cautions and warnings must be highly visible and must communicate very clearly to all segments of the audience. In addition, they must appear before the place in the procedure where the user can get into trouble. Often the phrasing, visual design, and even the placement on the product are specified by standards organizations such as ISO (the International Standards Organization).

Many kinds of helpful information can be treated as notes, and some notes are identified more specifically as tips and hints. Notes can appear almost anywhere in a procedure, and writers should choose the most useful location. Sakina warns her students against overusing notes and—especially—against adding multiple notes at the end of a procedure. Often the information in such notes really belongs in a step or feedback statement.

### Understanding How the Product Is Used

While writers must know how procedures are constructed, they also need a detailed understanding of how the product is used and, especially, what aspects of the product are apt to cause confusion and mistakes. When writers lack this knowledge (or if they are unskilled or unmotivated), their procedures tend not to give proper attention to what is complex and confusing. The writing may be fluent and the formatting may be attractive, but users will soon learn these procedures are not very useful. Sakina emphasizes how important it is to understand how the product is used.

### EXERCISES AND ASSIGNMENTS

Exercises and assignments are fundamental to instruction and learning. Here are two in-class exercises and one out-of-class assignment. Instructors will likely modify them to suit their own classes or use them as a springboard for their own exercises and assignments. For a slide deck and instructor materials that fully support a unit on procedures, go to https://faculty.washington.edu/farkas/FarkasInBridgeford/.

*Exercise: Writing a Lego Procedure and Observing the Users*

Sakina divides the class into groups. Each group consists of two teams with two to four members. She gives each team an object consisting of four arbitrarily assembled Lego® bricks. Each team also gets four loose bricks. Sakina uses the familiar "classic" bricks with 4 × 2 circular tabs. They can be purchased cheaply in bulk on the Internet.

Let's look at teams A and B of group 1. Keeping its object hidden from team B, team A spends about ten minutes writing a simple procedure—nothing but four steps—that will enable team B to duplicate team A's object. (The loose bricks can be helpful when writing the steps.) Team B does likewise. Visuals are not allowed and, in fact, are not necessary if the steps are written precisely.

Sakina offers some suggestions: describe the orientation of bricks using the points of the compass; specify that two (or more) bricks align vertically on certain edges or that they touch side by side; indicate that a brick is offset one (or more) tab rows from the edge of the brick beneath it. There are, in fact, various workable ways to write these steps. Sakina also suggests adding a limited amount of redundancy to steps (two ways of explaining how to position a Lego brick) as a safeguard against user error.

After the teams have completed their simplified procedure, team A gives its procedure to team B and receives team B's procedure. Taking about ten minutes, each team tries to duplicate the other team's object with its loose bricks using the procedure it was given. Then, the teams reveal the original objects, and the entire group collectively determines which of the two procedures succeeded or failed and what the problems were.

Sakina points out that Lego bricks are a mechanical system, but an unusually simple one. The highly constrained environment of identical snap-together bricks greatly simplifies the task of writing the procedure and the task of interpreting it. Sakina also makes sure the students don't miss a key point implicit in this exercise: without user testing, your chance of writing successful procedures goes way down.

If she wishes to extend the exercise, Sakina, in a fanciful, entertaining manner, informs the class that extraterrestrial invaders have begun exterminating Earthlings. Our only hope is to create millions of small objects that resemble the (tiny) spaceships used by another alien civilization the invaders are afraid of. So, we need procedures that will enable individuals and families across the world to save themselves by building one of these imitation spaceships and then placing it near their homes. Consulting freely with other students, each student now writes a procedure, which Sakina will review. To do so, each student revises their team's steps as necessary and adds a title and a brief introduction. One item of information the introduction should include is that the colors of the bricks do not matter to the aliens. Here is one way to explain how to build an imitation space ship:

*Build a Lego-brick object that will scare off extraterrestrial invaders*

To scare off the extraterrestrials who are attacking Earth, you must assemble four "classic" Lego bricks (the bricks with 4 × 2 circular tabs) to create a replica of a spaceship used by another alien civilization that the invaders greatly fear. Place your replica where it can be seen from the air near where you live or otherwise spend time. The color of the bricks does not matter. To protect your spaceship replica from the elements, you may want to glue the Legos into place and attach the completed object to a stable surface.

1. Set down a Lego brick (Lego 1) in a north-south orientation (or any orientation you designate as north-south).

2. Snap Lego 2 onto Lego 1 in an east-west orientation so its north and east edges align with the north and east edges of Lego 1.

3. Snap Lego 3 onto Lego 1 (not Lego 2) in a north-south orientation so its north edge touches the south edge of Lego 2. All three Lego bricks should now align on their east edges.

4. Snap Lego 4 onto Lego 3 in an east-west orientation so it is centered (cross-wise) on Lego 3. None of the edges of Lego 4 should align with any edge of Lego 3.

*Assignment: Writing a Procedure, Getting Feedback, and Revising*

Here is an out-of-class assignment that serves well as the main graded assignment for the course unit on procedures. Each student is assigned to write a (draft) procedure, illustrated or not, on the topic of their choice, excluding topics not sufficiently challenging. For example, a student might explain how to make a simple bicycle repair, apply eyeliner (quite difficult), carry out a computer task, and so on. Each student e-mails Sakina a brief description of their proposed topic so Sakina can approve it.

As part of the assignment, each student labels the procedure components and the kinds of steps that appear in the procedure. Each student
prints three copies of their draft procedure, submits one copy to the
instructor, and gives a copy to each of the two students with whom this
student will form an in-class working group. The group works round
robin: each student reads the procedure of the two other students in
their group and writes comments suggesting how the procedure could
be improved. (Visuals are treated only as draft versions that a professional graphic design would later polish.) Each student then reviews
the comments written by the other two students, asking for clarification
when necessary. Taking home the commented copies of their draft procedure, each student now revises this draft outside class and submits the
final version as a graded assignment.

The instructor can ask each student to submit the commented copies
of their draft along with the final version. Sakina can grade the assignments more thoroughly when she has the whole package: the draft,
the commented copies, and the final version. In addition, Sakina has
learned that if she tells the students in advance that she will be perusing
their comments, they do a better job of commenting. As with the Lego
exercise, students are getting user feedback on their procedures, but
here users do not attempt to perform the task.

*Exercise: Working in Teams to Make a Simulated Sports Video*

Instructors with some extra class time in their syllabus can use this lively
in-class exercise to give students hands-on experience coordinating
the numerous elements of a complex video procedure—a (simulated)
instructional sports video. The students work in teams with students
playing each of the following roles: a celebrity athlete, perhaps a golfer,
who demonstrates how to swing the club; the on-screen host, whose role
includes introducing and concluding the video; an off-screen narrator,
who explains the key details of the swing; a videographer, who makes a
circle of their hands to simulate a camera lens; and a director, who man-
ages the whole (chaotic) process.

Following a plan developed by the director and the team as a whole,
the videographer steps in and out of the action to visually represent
close-ups and medium shots. He also stiffens his fingers to indicate
stop motion at key moments in the golf swing and uses a slow, roll-
ing hand gesture to indicate slow motion. (The athlete freezes or
slows their swing in synch with the narration and the videographer's
gestures.) As the teams successively demo their simulated videos and

receive feedback from the instructor and the class, the sophistication of the demos improves.

## CONCLUSION: ADAPTING INSTRUCTION FOR DIFFERENT COURSES

Introductory technical communication courses differ greatly. Instructors with limited space in their syllabus or who are teaching less-experienced students can simplify the introductory concepts presented in this chapter. One way is to talk about text, graphics, video, and audio without introducing the concepts of medium and modality. To simplify further, instructors can bypass the introductory concepts altogether and base their instruction on the section "Teaching How to Write Procedures."

To simplify the section "Teaching How to Write Procedures," instructors, while retaining the key distinction between steps and feedback, can drop the distinctions about the kinds of steps. Students, however, will then be left on their own to intuit when to write the different kinds of steps. All the shortcuts described here are workable. However, a solid conceptual understanding of procedures ultimately results in better procedure writing and the ability to reason about good and bad design.

## DISCUSSION QUESTIONS

1. Instructors usually assign a textbook in their introductory technical and professional writing course, and these textbooks almost certainly include a chapter on writing procedures. In what ways is this inclusion a help and in what ways is it a problem? How do we best coordinate the information in this chapter with the chapter on procedures in our textbook?

2. Although there are important and obvious benefits in bringing professional technical writers into our classes, these practitioners often have a narrow perspective and speak as though the only way to write procedures is the way they do it at their company. How, then, can we choose the most promising guest instructors from industry, and how (and to what extent) can we prepare them to speak in our classes? Finally, is it discourteous to express some disagreement with the practitioner who served as your guest instructor during the previous class?

3. Various kinds of procedures—recipe books, computer manuals, instructional videos—can be regarded as specific genres, and these genres can

be readily divided into subgenres. Procedures generally might be considered as a broad genre or genre space. What are the benefits of using the concept of genre in teaching procedures? Can the idea of genre be a productive addition to the conceptual framework that begins this chapter?

4. Personas are descriptions of imaginary people who represent categories of users. Personas are primarily used in the area of user-interface design. Can personas be used for designing and writing procedures? Is it worthwhile to introduce personas in our classes?

5. When learning to write procedures, students should focus on writing procedures that work in a particular communication situation. However, large organizations, to ensure consistency in their procedures and to control costs, regularly develop database-driven content management systems that enable the documentation for a suite of related products to be spun off semiautomatically from a central repository of procedures and procedure components. Within the limitations of an introductory technical writing course, can we and should we give students a glimpse of the problem of reuse within an enterprise? How does reuse affect the planning and writing of procedures? What metadata might we associate with particular procedures and procedures components to enable efficient reuse?

## References

Agricola, Georgius. 1556. *De Re Metallica.* Translated by Herbert C. Hoover and Lou H. Hoover. London: Mining Magazine. Project Gutenberg. Released November 14, 2011.

Albers, Michael J. 2003. "Multidimensional Audience Analysis for Dynamic Information." *Journal of Technical Writing and Communication* 33 (3): 263–79. doi: 10.2190/6KJN-95Q V-JMD3-E5EE.

Alexander, Kara P. 2013. "The Usability of Print and Online Video Instructions." *Technical Communication Quarterly* 22 (3): 237–59. doi: 10.1080/10572252.2013.775628.

Carroll, John M., and Mary Beth Rosson. 1987. "The Paradox of the Active User." In *Interfacing Thought: Cognitive Aspects of Human-Computer Interaction,* edited by John M. Carroll, 80–111. Cambridge: MIT Press.

Coe, Mariana. 1996. *Human Factors for Technical Communicators.* Hoboken, NJ: Wiley and Sons.

Farkas, David K. 1999. "The Logical and Rhetorical Construction of Procedural Discourse." *Technical Communication* 46 (1):42–54.

Farkas, David. K. 2010. "The Diagnosis-Resolution Structure in Troubleshooting Procedures." In *Proceedings of the IPCC (International Professional Communication Conference),* Enschede, The Netherlands. doi: 10.1109/ipcc.2010.5529808.

Graham, S. Scott, and Brandon Whalen. 2008. "Mode, Medium, and Genre: A Case Study of Decisions in New-Media Design." *Journal of Business and Technical Communication* 22 (1): 65–91. doi: 10.1177/1050651907307709.

Jimenez-Crespo, Miguel A. 2013. *Translation and Web Localization.* Abingdon: Routledge.

Krull, Robert. 2001. "Writing for Bodies in Space." In *Proceedings of the IPCC (International Professional Communication Conference),* 173–82. Santa Fe, NM: IEEE. doi: 10.1109/IPC C.2001.971562.

Mayer, Richard, ed. 2005. *The Cambridge Handbook of Multimedia Learning.* Cambridge: Cambridge University Press. doi: 10.1017/CBO9780511816819.

van der Meij, Hans, and Mark Gellevij. 2004. "The Four Components of a Procedure." *IEEE Transactions on Professional Communication* 47 (1): 5–14. doi:10.1109/TPC .2004.824292.

# 9

# A PRIMER FOR TEACHING ETHICS IN PROFESSIONAL AND TECHNICAL COMMUNICATION

Paul Dombrowski

As a technical communication instructor, you know it is important to cultivate a sense of ethical responsibility in your students. You also know, though, that the topic of ethics can be a messy quagmire at any time but particularly so for students of technology used to thinking in terms of hard, indubitable facts and knowledge. This chapter will show you how important ethical awareness is in technical communication with real examples of how word choice and writerly perspective can have serious consequences. These examples will also keep you grounded in concrete specifics while avoiding the frustration of interminable debates about abstractions.

Ethics is an important element in all manner of human interaction, including professional and technical communication. There are, though, several reasons ethics is a particularly challenging topic for our profession. One reason is that by its very nature, as we will see, ethics *is* what is debatable and must be continually, consciously considered, weighed, and acted upon. It is not reducible to a set of tables or an algorithm. It is not measurable or weighable but yet is quite real, as examples of ethical lapses repeatedly bring to our attention. It cannot be given or awarded, though it can be cultivated, like bringing a seed to fruition within a person. Another reason is comparable to a common response when one is questioned, What is art? For ethics as for art, the response oftentimes is, I cannot say what it is, but I know it when I see it. Trying to foster a discussion of the ineffable can challenge even the best teacher, but the effort is worth it.

Another important reason discussing ethics is challenging is that the boundaries of ethics are amorphous, like a mist. Though philosophers over the years have sought to make ethics as clear and definite as possible, the nature of the matter is that all such efforts somehow fall short.

DOI: 10.7330/9781607326809.c009

In recent years, we find the concept of ethics being intermingled with morality, politics, and religion, such as Emmanuel Levinas's (1972) positing of ethical awareness as fundamental to our experience of "the other," or Michel Foucault's (1975) critique of the formidable but unseen power of discursive formations.

In the earlier days of professional and technical communication, ethics was generally understood rather narrowly and concretely as adherence to clarity, correctness, and precision in communication, irrespective of the content or larger context. Later, however, Greg Clark (1987, 1994), in two articles, Stephen Doheny-Farina (1989), in one article, Bruce W. Speck (1989), in one article, and Dale Sullivan (1990), in a single article, together set the stage in 1980 through 1990 for a more expansive view of ethics in technical communication, beyond correctness and precision. Later, Elizabeth Flynn (1997) and a host of other scholars such as Nel Noddings (2013) and Elizabeth O. Smith and Isabelle Thompson (Smith and Thompson 2002) highlighted the feminist dimensions of ethics and values in technical communication, while Lee Brasseur (1993) critiqued the ethics of science and technology themselves. There is also the commonplace notion of the Protestant work ethic to keep in mind, while Sam Dragga (1999) has written about Confucian ethics in technical and professional discourse. Brenton Faber (1999) has examined the role of "intuitive ethics" in professional discourse. His work aligns with Sam Dragga's (1996) revealing study of rudimentary ethical dilemmas on the job, which finds significant differences among professionals in the same field, but more important, finds also that the rationales for such decisions are often unclear and ad hoc. Dan Voss and Madelyn Flammia (Voss and Flammia 2007) have explained the intercultural breadth of technical discourse, implicitly calling for a view of ethics not entirely culture dependent. Within science itself, there is debate about the ethics of embryonic stem-cell research and cloning. And who would not recognize Marxism as a value system with a distinct ethic? In light of these issues, I consider the term *ethics* broadly and loosely, though not inappropriately, to mean a system of values with which one implicitly or explicitly identifies or enacts, sometimes with internal inconsistencies or external conflicts. One can be associated with various such systems depending on circumstances or conscience.

This chapter is a primer for teaching about ethics and professional and technical communication. It represents only one way in which instructors can introduce and explore the topic of ethics and values in technical and professional communication in the classroom, reflecting my own experiences. It is certainly not meant to be prescriptive, only suggestive. There are many other important researchers, theorists, critics, and professionals in the field whose names do not appear here. The absence is not meant as a slight but only reflects my concern as a writer for clarity and focus. After exploring some of the background of ethics, I examine a number of specific instances and documents that will help you foster discussions of ethics in your own courses.

My presentation here stems to some degree from my book *Ethics in Technical Communication* (Dombrowski 2000), which contains fuller development of several examples presented here as well as a general introduction to the theory and history of ethics studies. An approach to ethical considerations in technical communication rather different from my own is that of James E. Porter in *Rhetorical Ethics in Internetworked Writing* (Porter 1998). Still another approach you might consider is that of Lori Allen and Dan Voss in *Ethics in Technical Communication: Shades of Gray* (Allen and Voss 1997). Yet another is Mike Markel's (2000) *Ethics in Technical Communication: A Critique and a Synthesis*. Or see Richard L. Johannesen and Kathleen S. Valde's *Ethics in Human Communication* (Johannesen and Valde 2007) for a much broader perspective. Each has its individual approach to ethics, so all five merit being perused for their differences and their commonalities as they might resonate with your own interests. I urge you also to participate in workshops on applied and professional ethics if available at your institution to see how others grapple with the issue of trying to do the right thing in their fields.

## IN THE CLASSROOM

I teach a variety of technical communication courses, both undergraduate and graduate, and include ethics components in each course from the very start. In general, I do not distinguish between ethics and values or value systems.

Rhetoric is a subject addressed in another chapter of this book. I see ethics and rhetoric as intimately linked, like two sides of a single coin. Though Aristotle developed separate treatises on rhetoric and on ethics, he did not explicitly separate them in principle (Aristotle 1984, 2009). Indeed, he said rhetoric has to do with what is debatable or in doubt, and certainly that is the case for most ethical situations. We speak or write from a stance of values, whether consciously or unconsciously, such as corporate profitability or personal advancement. Those to whom we communicate themselves hold their own ethical principles or value systems, which can vary with the breadth of the audience. Plato in *Phaedrus* explains the role of goodness and ethics in distinguishing between noble and ignoble rhetoric (Plato 1999). Quintilian in his *Institutio Oratoria* expounded on an entire system of education whose aim was the development of "vir bonus, dicendi peritus," a good [that is, ethical] man speaking well. I find very useful Richard Weaver's explication of Plato's *Phaedrus* from his excellent book *Ethics of Rhetoric* (Weaver 1953), as well as Jurgen Habermas's *Moral Consciousness and Communicative Ethics* (Habermas 1986).

Though a few of the documents examined here were not published recently, that can be an asset. Events recent in the minds of students may be clouded or prejudged by ethical impressions already arrived at. Consider, for instance, waterboarding in the current context of international terrorism. Consider, too, the ethics of using unmanned combat aerial vehicles (UCAVs, or drones) to kill targeted people from control centers half a world away from the action. Some UCAVs are even designed to seek out targets and destroy them without commands, autonomously. Furthermore, though the ethical issues in the *Challenger* disaster were tragically replayed almost exactly in the more recent *Columbia* disaster, the sad fact of the repetition can be fully grasped only by understanding the first instance. The most important thing for our purposes is that the examples shown here are clear and compelling.

What follows are some of the modules I use in my own courses that could be easily adapted to your own needs.

### Blowing Smoke

This module has to do with the ethical paradox of telling a lie while telling the truth. It is also an example of the powerful role of the values of corporations and even industries, particularly profitability, in displacing conventional personalistic ethical values such as health and honesty. In class, we examine in detail an infamous advertisement run in all the major newspapers of the nation. Though it was published in 1954, what we learn from it is as relevant today as it was then—it is, in fact, an archetype of sophism in its traditional dishonorable sense. It involves manipulating public perceptions by alleging doubt or controversy. We will see these techniques played out again in two other modules.

The full-page ad titled *A Frank Statement to Cigarette Smokers* was placed in the *New York Times* and hundreds of other newspapers on January 4, 1954. I first explain the context of the times. Cigarettes were widely advertised in all manner of media then without any warnings and with much glamour. You can readily find numerous images from a web search of *cigarettes, 1950s,* and *ad.* The smiling faces of actors and actresses, doctors, and even Santa Claus urge smoking this or that brand of cigarettes. I also point out that at this time scientific researchers, including some funded by the federal government, were documenting the association between smoking and diseases such as cancer. These serious, scientific studies were perceived as a threat by the tobacco industry, which responded with redoubled advertising efforts, including this full-page public statement to the entire nation. The ad was underwritten with the signatures of fourteen executives of the largest tobacco companies in the United States.

In class, we first read the ad and discuss its overt message. We then go through the document line by line with an ethically critical eye. Students readily recognize the pervasive duplicity of the writing—technically true taken literally and denotatively at face value, but factually and semantically untrue and unethical in connotation and implication. Although stating that an interest in public health and welfare is paramount, the true message is entirely the opposite, serving the interests of the tobacco industry—it is as though white becomes black when understood from the point of view of the communicators. I reiterate that although this is an old document (see app. 9.A for links to this document), it played a prominent role in the much more recent legal suits of the 1990s and 2000s against the tobacco industry and in continuing suits.

I highlight here a few examples from this document rife with duplicity, pretense, and deception that would not be obvious to an uncritical or unwary reader. The first sentence speaks of "experiments with mice" and ends with the words "human beings," suggesting that the rigorously established findings were only experimental and not well established while further suggesting the possible inapplicability of these findings to humans, who are not the same as mice. Although the use of the word "theory" is technically true, its manner of presentation suggests it is mostly hypothetical and speculative rather than grounded in inferences from empirical observations. The words "in some way" echo this fallacious representation. Likewise, "not regarded as conclusive." Using the terms "disregarded" and "lightly dismissed" suggests the tobacco companies are being ethically cautious, while in reality, they are not.

Regarding the expression "eminent doctors and research scientists," I ask the class, what would it take to make this expression factually true? Supposing there exist across the nation ten thousand doctors and research scientists, it would take only four out of ten thousand to make the statement true. By not specifying numbers or proportions, the writers suggest to the naïve reader that the numbers are substantial while the reality is that the numbers are miniscule.

The fifth paragraph is particularly egregious. Stating that the public's health is a responsibility of greater concern for them (the industry) than anything else (implied in "paramount") is flatly false on the face of it. Industries and corporations by their very nature must make profitability and continued existence paramount above other considerations. We see this in our own times when it takes the legal force of the federal government to take corporations to task for their ethical lapses concerning public safety, such as in vehicle recalls for faulty airbags or faulty accelerators.

The key word in the sixth paragraph is "believe." The companies are not saying their products *are* not injurious to health—which would be a statement of fact—only that they "*believe*" their products are not injurious. Their statements might be true, but that would not make it true that cigarettes are not injurious. In the eighth paragraph, the writers state that the very suggestion or suspicion that cigarette smoking is injurious to health "is a matter of deep concern to us." This statement is factually true in that the signatories as representatives of impersonal, profit-driven corporations must be concerned that their profitability and even existence are being challenged, but the concern is not for the health of smokers.

To illustrate the depersonalized business frame of mind behind an ad of this sort and the appalling lack of ethical responsibility behind it, we examine a document written somewhat later, in 1969, for one of the signatories of *A Frank Statement*, the Brown and Williamson tobacco company. By then the context become even more challenging with the publication of the US Surgeon General's landmark report, *The Health Consequences of Smoking*. This combative and blunt document is commonly known by its most frightening statement, "Doubt is our Product" (*Doubt* 1969; see app. 9.A for a link to this document).

It outlines an action plan to oppose antismoking charges and activities presented by a marketing firm to the Brown and Williamson tobacco company. The document begins with its statement of purpose: "[To] make a proposal to you for a B&W [Brown and Williamson] project to counter the anti-cigarette forces." The paragraph ending page 4 and beginning page 5 merits quotation in full.

> *Doubt is our product. It is the best means of competing with the "body of fact" that exists in the mind of the general public.* It is also the means of establishing a controversy. . . . Doubt is also the limit of our product. Unfortunately we cannot take deposition directly opposing the anti-cigarette forces and say that cigarettes are a contributor to good health. *No information that we have supports such a claim* [emphasis mine]. (Doubt 1969; see app. 9.A for a link to this document)

This document, too, has played a significant role in the famous antitobacco suits of the 1990s and later.

We then view a YouTube video aptly titled, *The Seven Dwarves: Tobacco CEOs Testifying to Congress*, in 1994 (*The Seven* 1994; see app. 9.A for a link to this document). The video is important for its relative recency compared to *A Frank Statement* and for its video modality, which readily engages student interest. A member of the House Subcommittee on Health and the Environment, Representative Ron Wyden, pointedly and simply asks each of the seven testifying CEOs whether nicotine in cigarette smoke is addictive. Their brief simple responses speak volumes. Not one of them states whether nicotine is in fact addictive; instead, each one evades and obfuscates the issue by saying he does not "believe" nicotine is addictive. Because each one was aware of numerous scientific studies revealing nicotine is addictive, each would have perjured himself by stating it was not addictive. Instead, each in turn transmutes the question from one of scientific fact to one of personal belief. In this rhetorical transmutation, they sidestep their ethical responsibility to the public.

### Smoke and Mirrors

This module is a follow-up to the preceding section in bringing the same ethical matter into the present for students. It involves, as *Doubt* did, manipulating public perceptions by alleging and fabricating doubt or controversy. In this case, the doubt and controversy being manufactured has to do with global climate change. I usually present this module after the class views the video *An Inconvenient Truth* about global climate change (Gore and West 2006). In it, Al Gore describes the paradoxical situation of there being next to no opposition among peer-reviewed scientific journal articles about global climate change, while in the popular, nonscientific media, there are continual references to there being a debate or controversy as to whether global climate change is a factual reality. Discussion begins with the questions, how is it, and why (including value systems) is it, that scientists think one thing certainly and the public thinks the scientists are at odds among themselves on this matter?

The Union of Concerned Scientists is an international organization of tens of thousands of scientists of like ethical minds concerned about public policy and the uses to which scientific knowledge is put. In 2007, the Union of Concerned Scientists USA branch published a famous report titled quite descriptively *Smoke, Mirrors & Hot Air: How Exxon Mobil Uses Big Tobacco's Tactics to Manufacture Uncertainty on Climate Science* (*Smoke* 2007; see app. 9.A for a link to this document).

I ask the class to skim the lengthy report, then read closely the follow-ing items: "Executive Summary"; appendix A; appendix B, particularly table 3; appendix C, sections, 1998 memo; and the sample mark-up of the revised draft plan by Phil Cooney (also described and characterized in *An Inconvenient Truth*) (*Smoke* 2007; see app. 9.A for a link to this document).

We first discuss the content, aim, and implicit ethical system of the writers of the document. Then, as a writing assignment, I ask students to report on the following points. First, they summarize the overall argument and key points. Do the writers do an effective job of making a claim, arguing the reasons it is important and true, and providing ade-quate support or evidence to back up their claim convincingly? Describe, too, the role of "front" organizations in deflecting accountability from the actual source of a message and the intended effect of having a single message delivered through a number of disparate sources. Second, I ask them to tell me what they think about all this. (No correct or incor-rect answers here and no grade on this one.) Note that UCS USA does not claim that ExxonMobil or any of the agencies involved have done anything illegal, though UCS certainly strongly communicates that it is unethical and wrongful when the tobacco industry confuses the public and attacks well-founded science only so they can make more money and so people can smoke themselves to an early grave. Do you your-self, I ask, think it is ethical and honorable that ExxonMobil and other major profit-generating companies defend their activities by attacking those who report the true reality of climate change? After all, the entire industry of marketing and advertising revolves around advancing the interests of and creating profits for one's corporation, as we all know. Is the age-old caution *Caveat emptor!* (Buyer beware!) sufficient protection for the reader or consumer? Keep in mind that the very idea of "corpo-ration" is that of a nonhuman, impersonal entity deliberately separated from personal responsibility and liability—and personhood is where eth-ics abides. For further reading on this subject, I recommend the book *Merchants of Doubt* by Naomi Oreskes and Erik M. Conway (Oreskes and Conway 2011).

### Feynman's Dissent

This module involves a positive example of ethics in practice. It examines an important document relating to the space shuttle *Challenger* disaster in 1986. It reveals to students how highly technical subject matter—literally, rocket science—can be communicated with clarity and correctness but, most important, with a powerful sense of ethical responsibility. Indeed, aside from the document, Richard Feynman's dissenting opinion is itself an ethical act. The examination allows students to apply their own well-informed judgment in analyzing and evaluating the writing of someone else, in this case an accomplished scientific and technical expert. Ethics, after all, is to a large extent an inherently personal matter, and we all are already practicing ethical critics in our own lives.

I begin by presenting images of the shuttle crew and a video of the launch until disaster struck after seventy-one seconds. These images make vivid to students the actual persons involved along the ethical lines Sam Dragga and Daniel Voss recommend in order to humanize our technical discourse (Dragga and Voss 2001, 2003).

The commission investigating the disaster had an odd, ethically suspect composition. Its fourteen members included many individuals employed at one time or another by NASA or who closely worked with NASA, leaving the impression to some that NASA was investigating itself. There were only three scientists on the commission, one of whom, Richard Feynman, was a Nobel laureate in physics. Feynman was famous for his brilliant lectures at Caltech, for his development of quantum electrodynamics, and for his knack for making the abstruse comprehensible.

The report, spanning five volumes, can be found at numerous sites, including at NASA (Presidential Commission 1986). When it came time to sign the commission's report and submit it to the president, Feynman balked. He disagreed with the overall tenor and substance of the report but was permitted only to submit his own separate statement, "Personal Observations," which was relegated to appendix F of volume 2, where it was not likely to be paid much attention (Feynman 1986; see app. 9.A for a link to this document).

The primary learning goal of this module is to see how values and ethics can play a substantial role in the way particular people write about their technical subjects, even rocket science. It also shows, similarly, how a subject supposedly as concrete and factual as rocket science can be dramatically shaped by social context and therefore by whatever value system pervades that context. I emphasize the importance of personal voice and a sense of ethical responsibility, as they are reflected in technical communication. Students are asked to read only the introduction, the section titled "Solid Rockets (SRB)," and the conclusion of Feynman's "Personal Observations," which are not overly technical (Feynman 1986; see app. 9.A for a link to this document).

Students then write about what they learned about ethical responsibility from the language of this technical document and about the powerful influence of social context on the understanding of technical matters. The difference between managers and engineers is striking, and the metaphorical representation of the O-ring charring exemplifies Feynman's masterful communication abilities. I point out that ethics and values are often thought of as a personal matter, just a matter of opinion, and ask them whether Feynman thinks this way. The reading is somewhat technical, so you might want to urge students not to get bogged down in the details. It really does not take a rocket scientist to follow the general thrust of Feynman's argument and conclusions. The key point I hope students take from this exercise is the crucial role that personal voice and personal ethical responsibility can take in even the most complex of technical environments.

## TWO ACTUAL "SMOKING-GUN" MEMORANDA

This module focuses on two actual technical memoranda famous for how they illustrate the ethical impact of style and organization. They are "smoking-gun" documents in that they represent a crux, a point of decision about a course of action. In the one case, the writer failed to make his points as clear, significant, and persuasive as they reasonably could have been. That the needed action did not in fact occur is thus at least in part the fault of the writer rather than of the reader. In the other case, the writer clearly took earnest steps to make his writing as clear and emphatic as possible in order to try to precipitate much-needed action. That the needed action did not in fact occur is thus the fault of the readers and not of the writer. Students readily recognize the stark differences between the two pieces.

Both of these real memoranda have been cited in numerous governmental investigations as key technical reports having to do with very serious technical disasters. I remind the class that professional and technical communications should be crafted carefully to make matters as clear and easily understood as possible to all readers. The learning objective is achieved by contrasting the two and considering the deliberate choices made by the writer in each case.

### Three Mile Island

I start by contextualizing the two documents. The first is an internal memorandum within Babcock and Wilcox, Inc., the manufacturer of the nuclear reactors and the power generation equipment at Three Mile Island in Pennsylvania (Staff of the President's Commission 1979; see app. 9.A for a link to this document). The Three Mile Island accident on March 28, 1979, is the biggest nuclear-power accident in US history as well as the most expensive and the most highly publicized. Though there was no complete meltdown, it came close. Fortunately, there was little actual leakage of radiation into the environment, though, again, the possibility for much worse was real and close.

I point out that one generally need not be a nuclear engineer to catch the drift of this piece or at least its subject and purpose. Students are asked in class to answer the following questions and discuss whether they can readily be answered from the document.

- What is the key phrase in this document, and where does it occur? Does its placement ensure it will be detected and given full attention?
- Is there any indication of danger, urgency, or safety issues in its language or organization?
- Is there any sense of a personal voice or ethical responsibility communicated in its style?

The discussion usually reveals confusion and unclear focus. I explain that all the content in this memorandum is entirely true and correct but lacks a number of features that could have helped bring about appropriate attention to address the threat of possible meltdown. This memorandum and several others were highlighted in the *Report to the President's Commission on the Accident at Three Mile Island* (Staff of the President's Commission 1979; see app. 9.A for a link to this document) as "smoking guns." The full report traces a long history of friction between Babcock and Wilcox, manufacturer of the reactor system, and Metropolitan Edison Company, the operator of the plant. This friction included a large number of remarkably convoluted, turbid, and even hostile communications between the two parties.

There are many things not right about this piece of writing that make it ethically problematic. I emphasize to the class that the next memorandum we will look at is very different and much better, which shows that the problem is not the technicality of the subject matter but the ethical care of the writer. Among the points I highlight are these:

- The subject line has no language of danger or urgency.
- The first paragraph likewise has no language of danger, which is precisely where such language should be placed prominently to ensure the readers' attention and action.
- The first paragraph speaks of two other documents, almost a diversion from urgent attention to the actual document before the reader. It also speaks of "philosophy," language that puts the subject on a plane of theory or abstraction rather than of concrete reality.

- It raises two important rhetorical questions but fails to offer possible answers to those questions, which could have indicated the gravity of this issue and of this communication.
- The next-to-last paragraph contains the key phrase "uncovering the core" (usually less than half the class catches it). This key phrase is buried in the middle of a paragraph, the beginning and ending sentences of which give no indication of danger or urgency. "Possibility of Uncovering the Core" would have been an excellent subject line and opener to the first paragraph.
- The last paragraph is trite and again does not communicate a sense of danger or urgency.

### Morton Thiokol and the Shuttle Challenger

The next memorandum is a very different one. It clearly illustrates a style and organization distinct from the Three Mile Island memorandum, yet both include highly technical information. The memorandum addresses the ongoing problem of charring and leakage in the O-ring seals between the segments of the huge solid rocket motors on the space shuttle (two, one on each side of the shuttle and main tank, each about 12 feet in diameter and 125 feet long). They are like enormous Fourth of July skyrockets. Any leakage through the seals would burn through the side of the booster and cause it to explode. Just six months after this memorandum was written, that is exactly what happened to the shuttle *Challenger* on January 28, 1986, with results we know all too well (see app. 9.A for a link to this document).

You do not have to be a rocket scientist to read and understand this piece of technical communication. Morton Thiokol Inc. built and fueled the two solid rocket motors for each shuttle launch. We discuss this memorandum in class, focusing on these questions:

- Is urgency, danger, or concern expressed clearly? How and where?
- Is there a sense of personal voice and personal ethical responsibility? How and where?
- Are key points emphasized in obvious places and reiterated? What words or phrases act as intensifiers or modifiers to heighten the sense of urgency and concern?

There are many aspects right about this piece of writing; I touch on only the most significant and striking, especially in contrast to the preceding memorandum.

- The subject line includes the words "failure" and "criticality," which would obviously immediately evoke attention from the reader as a serious matter.

- The first paragraph clearly states its purpose and uses ominous terms "to insure that management is fully aware of the seriousness" of the subject. The word "seriousness" itself in the first sentence states the gravity of the issue, reinforced by the word "problem" in the same sentence. Note how the sentence would read without "fully," "seriousness," and "problem," technically true but weak.
- In the second paragraph, the word "mistakenly" at the beginning of the paragraph makes emphatically clear that the prevailing mindset is incorrect. Note, too, how the paragraph would read without the words "mistakenly" and "drastically."
- In the third paragraph, the colloquial phrase "jump ball" stands out in this piece of technical distance to make sure the reader gets the message. The paragraph ends with stunning clarity and ethical force with "human life" as its final words. Note that the entire final sentence could have been left off for an adequate paragraph.
- The final paragraph is remarkable for its powerful personal and ethical voice in "my honest and very real fear," a voice rarely heard in communication about applied mechanics, author Roger Boisjoly's field, so is very likely to grab the attention of the reader.[1] The word "jeopardy," likewise unusual in technical discourse, emphasizes seriousness again. The entire final clause could not have been stated more clearly, effectively, or passionately.

In this smoking-gun case, the writer has done everything he could to communicate clearly and ethically to his readers. That no effective action took place in response to this communication was then the ethical responsibility of the readers.

Generally, students are engrossed in this assignment for its awful sense, in the case of *Challenger*, of a disastrous loss of human life that could have and should have been prevented. They write a report on each of the two smoking-gun memoranda, highlighting their differences and explaining the sense of ethical responsibility the writers might have felt.

As a final note, Roger Boisjoly and his team members strenuously argued against the launch of the shuttle *Challenger*. Not long after the two governmental investigations (presidential commission and congressional committee) were completed, Boisjoly lost his job at Morton Thiokol. He filed suit successfully under federal whistleblowing laws and was reinstated. Unfortunately, he was relocated to an obscure physical location and his activities were relegated to trivial tasks; in frustration, he resigned. He worked tirelessly to cultivate a serious awareness of the role of ethics in the engineering profession until his death on January 6, 2012. Read his article on the disaster and about him at the Online Ethics Center (Boisjoly 1989; see app. 9.A for a link to this document).

*Shuttle* Columbia: *Ethics in a Broader Sense*

This module is a follow-up to our examination of the *Challenger* memorandum. The points of connection are not just the shuttle per se but also the continuation of ethically problematic contextual forces that were shaping communication as well. Further, it is an expansion of the concept of ethics beyond personal decisions or particular instances to encompass the prevailing culture at NASA itself. The module does not reveal a particular smoking-gun document except for the findings of the Columbia Accident Investigation Board (CAIB) as reported in their entire document (which in itself is an exemplary piece of technical communication) (see app. 9.A for links to the *Report* 2003 and "Chapter 6" 2003). The CAIB report amounts to an ethical condemnation of the organizational culture of NASA itself. The chief learning objective is to understand how organizational circumstances can present profound ethical challenges for individual persons, with disastrous effect. Although it is about the cultural work ethic of NASA, the circumstantial forces it reveals are not at all particular to NASA.

Rather than requiring the class to read the whole report, I ask them to read instead an article written about both the *Challenger* and the *Columbia* disasters and the reports about them (Dombrowski 2006). It describes the crucial role of organizational context/culture in influencing discourse as discovered and reported by the CAIB. The report itself is remarkable in highlighting the powerful and pervasive role of context and organizational culture in shaping events in communications; the word "culture" appears many times throughout the document (I stopped counting at 125).

As with the *Challenger*/Morton Thiokol memorandum, I introduce the assignment with the group photo of the *Columbia* crew just before launch in order to humanize the technical subject matter. I then describe the disaster itself, which occurred on January 16, 2003 (within the memory of my students, many of whom live not far from the Kennedy Space Center). During the launch, a piece of insulating foam about the size of a briefcase broke off from the shuttle's huge tank as it was speeding at about two thousand miles per hour, striking the leading edge of the left wing and blasting a hole in it about the size of a laptop computer. Days later, the crew was returning to Earth when, unbeknownst to the crew, the hole in the wing allowed the hot frictional gases of reentry to penetrate the wing. The wing then collapsed and the vehicle disintegrated at about fifteen thousand miles per hour and about forty miles altitude. The crew perished.

To vividly demonstrate the real influence of cultural context on ethical issues of truth and clarity, I present from the report visuals taken from space of the "dings" to the tiles on the shuttle underside and several key pages from the document (Columbia Accident Investigation Board 2003; see app. 9.A for a link to this document). The chart shows the number of dings (dings defined as damage greater than one inch in diameter to the tiles on the underside of the shuttle) per shuttle mission number. Though the numbers indicated by the height of the columns is important, much more important is the bare fact that the Tile Protection System (TPS) included the original design requirement that these tiles should never be impacted by anything at any time. The reality of continual tile damage came to be accepted as normal and not a threat to the safety of the mission. This is a clear illustration of the social/cultural phenomenon of "the normalization of deviance" discussed in detail in the official reports of both the *Challenger* and the *Columbia* disasters.

The report blames as causative factors large-scale societal forces external to NASA, such as economic and political factors like unrealistic expectations for the system and reduced fiscal support over the years. More important, it also blames forces within the NASA culture, which we focus on in class. We discuss sections of the article, which follow the argument of the CAIB report itself:

- the specific design requirements that were defeated through the normalization of deviance;
- the roles of societal expectations, political expectations, and budgetary constraints;
- the reversal of assumptions—a catch-22 dilemma showing how efforts to determine the damage to the foam strike were undone by a confused chain of authority, an unwarranted exercise of authority, and confusion about the meaning of a particular phrase in an unanticipated situation;
- the normalization of deviance, a phenomenon not peculiar to NASA, and how it carried over from *Challenger* disaster, as though the lesson had not been learned.

Students report on what they learned about the invisible but real influence of cultural context on professional and technical communication, which can yield ethically problematic decisions and actions, and ultimately disaster.

## CONCLUSION

This chapter has taken a view of ethics broader than what is simply true, direct, precise, and depersonalized. This view assumes we are all ethicists already and inescapably in our responsibility to be mindful of the ethical issues our professional and technical communications might present. Sometimes, such issues can be anticipated in a code of ethics, but often such codes in their reductiveness and generality cannot offer specific guidance. It is an innate part of our human nature, particularly in our discourse among fellow humans, that we must face our ethical responsibilities whenever doubtful or problematic situations arise. I hope this primer will help you foster an active sense of ethical responsibility in your technical and professional communication students.

## DISCUSSION QUESTIONS

1. Examine several codes of ethics (e.g., Society for Technical Communication, Association of Teachers of Technical Writing). Discuss their limitations in providing guidance for particular professional and technical communication situations you can imagine.

2. Read the brief *New York Times* article "We Have Met the Enemy and He Is PowerPoint" about the overzealous use of PowerPoint to display complex information at http://www.nytimes.com/2010/04/27/world /27powerpoint.html?_r=0. Alternatively, ask students to find articles, especially related to technology, about the way PowerPoint can lead to leaving out key information and closing off discussion of fine points. Discuss the ethical implications of each misuse.

3. The Columbia Accident Investigation Board identified the social phenomenon of "culture" as a key causative factor in the disaster. Ethically, this conclusion seems to infer that the social organization has its own ethical responsibility separate and different from individual personalistic ethical responsibility. Ask the class to offer possible examples from the news or from their own lives. Discuss whether this social form of ethics/value system can or should override personalistic responsibility. The Nazi program leading to the Holocaust is a ready instance for such discussion. Other instances might be the suppression of knowledge about Takata air bags or the Volkswagen doctoring of engine software to yield false but desirable mileage readings.

## APPENDIX 9.A

Listed by module headings are web sources for items described in the text.

### BLOWING SMOKE:

*A Frank Statement*

> http://www.sourcewatch.org/index.php/The_Frank_Statement
> https://en.wikipedia.org/wiki/A_Frank_Statement
> http://archive.tobacco.org/History/540104frank.html
> https://www.industrydocumentslibrary.ucsf.edu/tobacco/docs
>   /#id=ltln0082

*The Seven Dwarves (Tobacco Company CEOs Testifying to Congress)*
> https://www.c-span.org/video/?c3336890/1994-ceos-testify. Accessed: December 22, 2016: Copy and paste into your browser.
> http://www.jeffreywigand.com/7ceos.php. Accessed December 22, 2016.
> *Doubt Is Our Product.* 1969. From Brown and Williamson tobacco company. See http://legacy.library.ucsf.edu/tid/rgy93f00/pdf (the document without context or commentary) or http://tobaccodocuments.org/land man/332506.html (a very detailed archive of key tobacco documents with commentary and cross-references). Accessed December 10, 2015.

*Smoke and Mirrors*
> *Smoke, Mirrors & Hot Air* by the Union of Concerned Scientists USA. 2007: http://www.ucsusa.org/assets/documents/global_warming /exxon_report.pdf.

*Feynman*
> Presidential Commission on the Space Shuttle Challenger Accident. 1986. *Report of the Presidential Commission on the Space Shuttle Challenger Accident.* http://history.nasa.gov/rogersrep/genindex.htm. Accessed December 12, 2015.
> Feynman, R. P. "Personal Observations on the Reliability of the Shuttle." 1986. In *Report of the Presidential Commission on the Space Shuttle Challenger Accident.* Vol. 2, Appendix F. http://science.ksc.nasa.gov/sh uttle/missions/51-l/docs/rogers-commission/Appendix-F.txt (read as "51-lower case L").

*Two Smoking Guns*
Three Mile Island
> Staff of the President's Commission on the Accident at Three Mile Island. 1979. *Report to the President's Commission on the Accident at Three Mile Island.*

The entire report is available here and elsewhere:

> http://babel.hathitrust.org/cgi/pt?id=purl.32754062964238;view=1up;seq=5

Within the report, the memorandum can be found as appendix N on p. 227, titled memorandum from D. F. Hallman to B. A. Karrasch, August 3, 1978.

Also available as the memorandum itself at http://www.technical communicationcenter.com/2009/10/14/bad-technical-writing-can-lead -to-disaster/ and at https://web.njit.edu/~lynchr/eng624/TMI-MEMO .html

### Morton Thiokol and Space Shuttle Challenger

*Report to the President by the Presidential Commission on the Space Shuttle Challenger Accident* (commonly known as the Rogers Commission report). 1986 Morton Thiokol Interoffice memo, 31 July 1987. From R. M. Boisjoly to R. K. Lund. Found at volume 1, p. 249: http://his tory.nasa.gov/rogersrep/v1p249.htm and at volume 4, p. 807: http://history.nasa.gov/rogersrep/v4p807.htm. Accessed December 14, 2015.

"Roger Boisjoly—The Challenger Disaster." Online Ethics Center. Last updated August 29, 2016. http://www.onlineethics.org/cms/7123 .aspx.

"Ethical Decisions-Morton Thiokol and the Space Shuttle Challenger Disaster–Index." http://www.onlineethics.org/Resources/thiokol-shuttle.aspx.

Boisjoly's own account of events leading up to the disaster and subsequent events relating to his career. Accessed December 15, 2015.

Online Ethics Center for Engineering and Science (OEC). http://www .onlineethics.org/

### Columbia

Columbia Accident Investigation Board. 2003. *Report of the Columbia Accident Investigation Board*, Chapter 6. http://history.nasa.gov/colum bia/Troxell/Columbia%20Web%20Site/CAIB/CAIB%20Website/CA IB%20Report/Volume%201/Part%202/chapter6.pdf.

Figure 6.1–6, from volume I, chapter 6, p. 127, shows the number of "dings" per shuttle mission.

### Note

1.  Roger Boisjoly was a mechanical engineer and aerodynamicist working for Morton Thiokol, Inc., which built the Solid Rocket Boosters for the space shuttle. He is famous, with other colleagues, for strongly and repeatedly opposing to management the launch of shuttle *Challenger* on the night before the launch but was overruled. For his conscientious and brave efforts, he was awarded the Award for Scientific Freedom and Responsibility by the American Association for the Advancement of Science in 1988.

*References*

Allen, Lori, and Daniel Voss. 1997. *Ethics in Technical Communication: Shades of Gray*. Hoboken, NJ: Wiley.

Aristotle. 1984. *The Rhetoric and the Poetics of Aristotle*. Translated by W. Rhys Roberts and Ingram Bywater. New York: Modern Library.

Aristotle. 2009. *The Nicomachean Ethics*. Translated by David Ross, edited by Leslie Brown. Oxford: Oxford University Press.

Brasseur, Lee. 1993. "Contesting the Objectivist Paradigm: Gender Issues in the Technical and Professional Communication Classroom." *IEEE Transactions on Professional Communication* 36 (3): 114–23.

Clark, Greg. 1987. "Ethics in Technical Communication: A Rhetorical Approach." *IEEE Transactions on Professional Communication* 30 (3): 190–95.

Clark, Greg. 1994. "Professional Ethics from an Academic Perspective." *Journal of Computer Documentation* 18 (3): 32–8. Retrospective essay invited for an issue featuring a reprint of his "Ethics in Technical Communication: A Rhetorical Approach" (1987) and four essays in response.

Columbia Accident Investigation Board. 2003. *Report of the Columbia Accident Investigation Board*. Vols. 1–4. Washington, DC: NASA and US GPO.

Doheny-Farina, Stephen. 1989. "Ethics: A Bridge for Studying the Social Contexts of Professional Communication." *Journal of Business and Technical Communication* 3 (3): 70–88.

Dombrowski, Paul M. 2000. *Ethics in Technical Communication*. Boston, MA: Allyn and Bacon.

Dombrowski, Paul M. 2006. "The Two Shuttle Accident Reports: Context and Culture in Technical Communication." *Journal of Technical Writing and Communication* 36 (3): 231–52.

Dragga, Sam. 1996. "'Is This Ethical?': A Survey of Opinion on Principles and Practices of Document Design." *Technical Communication* 43 (3): 255–65.

Dragga, Sam. 1999. "Ethical Intercultural Technical Communication: Looking through the Lens of Confucian Ethics." *Technical Communication Quarterly* 8 (4): 365–82.

Dragga, Sam, and Daniel Voss. 2001. "Cruel Pies: The Inhumanity of Technical Illustrations." *Technical Communication* 48 (3): 265–74.

Dragga, Sam, and Daniel Voss. 2003. "Hiding Humanity: Verbal and Visual Ethics in Accident Reports." *Technical Communication* 50 (1): 61–82.

Faber, Brent. 1999. "Intuitive Ethics: Understanding and Critiquing the Role of Intuition in Ethical Decisions." *Technical Communication Quarterly* 8 (2): 189–96.

Flynn, Elizabeth A. 1997. "Emergent Feminist Technical Communication." *Technical Communication Quarterly* 6 (3): 313–20.

Foucault, Michel. 1975. *Discipline and Punish: The Birth of the Prison*. Paris: Gallimard.

Gore, Al, and Billy West. 2006. *An Inconvenient Truth*. Directed by Davis Guggenheim. Hollywood, CA: Paramount.

Habermas, Jurgen. 2001. *Moral Consciousness and Communicative Action*. Reprint edition. Cambridge: MIT Press.

Herndl, Carl G., Barbara A. Fennell, and Carolyn R. Miller. 1991. "Understanding Failures in Organizational Discourse: The Accident at Three Mile Island and the Shuttle Challenger Disaster." In *Textual Dynamics of the Profession: Historical and Contemporary Studies of Writing in Professional Communities*, edited by Charles Bazerman and James Paradis. Madison: University of Wisconsin Press.

Johannesen, Richard. L., and Kathleen S. Valde. 2007. *Ethics in Human Communication*. 6th ed. Long Grove, IL: Waveland.

Levinas, Emmanuel. 2003. *Humanism of the Other (Humanism de l'autre homme)*. Translated by Nidra Poller. Urbana: Illinois University Press.

Markel, Mike. 2000. *Ethics in Technical Communication: A Critique and a Synthesis*. ATTW Contemporary Studies in Technical Communication. Santa Barbara, CA: Praeger.

National Academy of Engineering, Center for Engineering Ethics and Society (CEES). Accessed March 20, 2016.

Noddings, Nel. 2013. *Caring: A Relational Approach to Ethics and Moral Education*. 2nd ed. Oakland: University of California Press.

Oreskes, Naomi, and Erik M. Conway. 2011. *Merchants of Doubt: How a Handful of Scientists Obscured the Truth on Issues from Tobacco Smoke to Global Warming*. Reprint edition. London: Bloomsbury.

Plato. 1999. *Phaedrus*. In *Euthyphro, Apology, Crito, Phaedo, Phaedrus*. Reprint edition. Translated by Harold North Fowler. Loeb Classical Library. Cambridge, MA: Harvard University Press.

Porter, James E. 1998. *Rhetorical Ethics and Internetworked Writing*. Greenwich, CT: Ablex.

Presidential Commission on the Space Shuttle Challenger Accident. 1986. *Report of the Presidential Commission on the Space Shuttle Challenger Accident*. Washington, DC: US Government Printing Office.

Smith, Elizabeth O., and Isabelle Thompson. 2002. "Feminist Theory in Technical Communication: Making Knowledge Claims Visible." *Journal of Business and Technical Communication* 16 (4): 441–77.

Speck, Bruce W. 1989. "Ethics and Technical Communication." In *Technical and Business Communication: Bibliographic Essays for Teachers and Corporate Trainers*, edited by Charles H. Sides. Urbana, IL: NCTE.

Staff of the President's Commission on the Accident at Three Mile Island. 1979. *Report to the President's Commission on the Accident at Three Mile Island*. Washington, DC: US Government Printing Office.

Sullivan, Dale. 1990. "Political Ethical Implications of Defining Technical Communication as a Practice." *Journal of Advanced Composition* 10 (2): 375–386. Reprinted in *Humanistic Aspects of Technical Communication*. 1994. Edited by Paul M. Dombrowski. Amityville, NY: Baywood. Also reprinted in *Central Works in Technical Communication*. 2004. Edited by Johndan Johnson-Eilola and Stuart Selber. Oxford: Oxford University Press.

Voss, Daniel, and Madelyn Flammia. 2007. "Ethical and Intercultural Challenges for Technical Communicators and Managers in a Shrinking Global Marketplace." *Technical Communication* 54 (1): 72–8.

Weaver, Richard M. 1953. "The Phaedrus and the Nature of Rhetoric." In *The Ethics of Rhetoric*, by Richard M. Weaver. Chicago, IL: Regnery.

# 10

# WHAT DO INSTRUCTORS NEED TO KNOW ABOUT TEACHING COLLABORATION?

Peter S. England and Pam Estes Brewer

As a new instructor of technical communication, you have many choices available to you in creating instruction. You may be starting with a well-outlined course used by other instructors in your department, or you may be working on your own creating a course from scratch. Regardless, you have many choices ahead of you and limited time. How you implement (or don't implement) collaboration greatly affects the effectiveness of your technical writing course.

Effective use of collaboration supports rich and authentic contexts where students are more engaged and where their learning accurately reflects workplace contexts. Ineffective use of collaboration can leave both you and the students feeling as if significant learning has somehow eluded you. Most of us can think of classroom collaborations in which the teacher implemented collaboration ineffectively and students were left feeling as if they were thrown into a pool where their efforts were not perceived as any different from the other students' work even if they were contributing most of the work. Frustration and loss of learning opportunities resulted.

Teaching effectively with collaborative projects is challenging but well worth the effort. This chapter will help you start using collaboration deliberately and effectively in your classroom (whether it's face to face or online). And as we move from the *what* to the *why* to the *how* of collaboration, the arguments we present in this chapter are the same ones you should use with students to help them understand the importance of collaborating effectively and the types of learning available to them.

One of the biggest differences between the classroom and the workplace is in how writing gets done. In the classroom, writing is often an individual activity: a student writes a paper and receives feedback in the

DOI: 10.7330/9781607326809.c010

form of suggestions and a grade. The purpose of introductory technical writing courses, however, is usually "to better prepare students for the writing they will do on the job," and those same intro classes are usually the only exposure to technical communication for those not already majoring in technical communication (Meloncon and England 2011, 398). In the workplace, writing in a group is common. In many workplaces, such as engineering firms, collaborative writing is the norm rather than the exception.

As a new teacher of technical communication, with a goal of preparing students to collaborate effectively in a variety of workplaces, you might be familiar with a scenario in engineering such as building a hospital. Designing and building a hospital is even more challenging than other engineering projects because hospitals have special needs other buildings don't (MRI machines, backup power sources, and so on). In other words, a hospital has so many design aspects that a single person cannot hope to remember or account for everything. Or, you may have read complex documents in another field and considered the need for successful collaboration in creating those documents. It follows, then, that some parts of your course must be given to preparing students to write collaboratively in realistic, complex contexts. How can you effectively design instructional projects that help learners collaborate effectively when writing? Such projects take planning, with clear performance objectives and assessment for collaboration.

This chapter can help you with planning and using collaboration effectively in your technical communication classroom. We, the authors of this chapter, use collaborative activities regularly, including group presentations and writing assignments.

In this chapter, we examine the meaning of collaboration and its impact on learning. We then present recommendations for using, managing, and assessing collaboration in the technical communication classroom, and, finally, we provide samples of both face-to-face and virtual collaboration projects.

## WHAT IS EFFECTIVE COLLABORATION?

Defining *collaboration* is as difficult as defining words like *culture* or *rhetoric*. The concept of collaboration is complex and may vary in definition. It is useful to look at some of these definitions to help you develop a model of collaboration that works for you.

Lev Vygotsky (1978) viewed collaboration as a social construct wherein meaning is constructed during team interaction. Rebecca E.

Burnett, L. Andrew Cooper, and Candace A. Welhausen define collaboration as "an intentional, sustained interaction toward a common goal" (Burnett, Cooper, and Welhausen 2013, 54).

Marjorie Davis (2015) writes that the "best collaboration generates new ideas, expands and clarifies the problem, and can reveal that your real goal might even be something different." Beth L. Hewett, Charlotte Robidoux, and David Remley write that "[it] involves strategic and generative interactivity among individuals seeking to achieve a common goal, such as problem solving, knowledge sharing, and advancing discovery" (Hewett, Robidoux, and Remley 2010, 9).

Scholars also describe what collaboration is NOT, although they don't always agree. For example, Burnett, Cooper, and Welhausen (2013) write that collaboration is not

1. having group members write their sections separately and then put them together at the end;
2. having group members present only their segment of material within a project, and;
3. having one group member do all the work.

True collaboration results in outputs better than what could have been achieved by a single person.

## HOW DOES COLLABORATION AID LEARNING?

Besides realistically preparing students for a collaborative workplace, class collaboration supports effective learning. Although many books and articles have been published on collaboration and learning, we favor the ten principles set forth in 1998 by the Joint Task Force on Student Learning created by the American Association of Higher Education, the American College Personnel Association, and the National Association of Student Personnel Administrators. These principles have been widely applied over the years and are based on a critical review of seminal publications on student learning. Principles one and five point specifically to the importance of collaboration in the learning process:

- Learning is fundamentally about making and maintaining connections: biologically through neural networks; mentally among concepts, ideas, and meanings; and experientially through interaction between the mind and the environment, self and other, generality and context, deliberation and action.

- Learning is done by individuals who are intrinsically tied to others as social beings, interacting as competitors or collaborators, constraining or supporting the learning process, and able to enhance learning through cooperation and sharing. (Eison 2002, 160)

These two principles emphasize the importance of making connections among self and others as an essential element of learning rather than merely a desirable tool. In sum, effective collaboration is a critical classroom tool, helping students engage in learning and preparing them for the knowledge workplace.

## HOW DOES COLLABORATION PREPARE STUDENTS FOR THE WORKPLACE?

Incorporating collaboration effectively into the college classroom better prepares students for the knowledge work that lies ahead of them. However, effective collaboration is hard work and requires much more than giving students an assignment and expecting them to identify and navigate the nuances of effective collaboration. Ram Nidumolu, Jib Ellison, John Whalen, and Erin Billman note a "growing awareness of the critical need for improved collaboration" and that "countless efforts by companies to work together to tackle the most complex challenges facing our world today—including climate change, resource depletion, and ecosystem loss—have failed because of competitive self-interest, a lack of a fully shared purpose, and a shortage of trust" (Nidumolu et al. 2014, 77). On a more positive note, companies and products that are household names, such as Linux, Apache, and Mozilla, are the result of far-flung collaborative efforts of a multitude of people (Mudge 2015). Bob Mudge, president of Consumer and Mass Business at Verizon, writes that "collaboration is no longer just a strategy: it is the key to long-term business success and competitiveness. Businesses that realize this sooner rather than later will be the ones who win the game and succeed in the new global economy" (Mudge 2015).

Beyond traditional face-to-face collaboration, virtual worlds are used as a collaboration medium (Bosch-Sijtsema and Sivunen 2013; Brewer et al. 2015) for small team meetings, training, community building, and conferences. In addition, information communication technologies of all kinds (e-mail, videoconferencing, instant messaging, texting) are being used to support collaboration at a distance.

## MANAGING COLLABORATION IN THE CLASSROOM

In this section, we focus on ideas for incorporating collaboration into your classroom. First, we should mention some potential obstacles instructors may face when working collaboration into the curriculum.

Although some schools embrace a collaborative approach to education, you may find yourself working within an educational paradigm that privileges individual over collaborative learning. New assignments of any type will need to be tested and tweaked. We suggest instructors read their school's guidelines for course development and be ready to justify collaborative work to administrators as well as students. Although collaborative work is inherently practical in the workplace and likely to be a crucial part of any professional career, working in, and evaluating, groups of students rather than individuals may push at the boundaries of what is accepted in your unique teaching environment. Instructors wishing to increase collaborative work in their courses may find themselves practicing collaborative work strategies in getting new assignments approved and implemented.

Collaborative activities can take many forms, such as written, verbal, face to face, or online. All collaborative assignments have the potential to disrupt institutional practices and can sometimes lead to student dissatisfaction. In this section, we provide you with guidelines for implementing collaborative work effectively. You can apply these guidelines to your own classroom, regardless of how your student population might differ from ours.

### Create Buy-in

Instructors should be prepared to persuade administrators, students, and other instructors of the need for and potential benefits of collaborative versus individual work. In business, we talk about buy-in: that is, the extent to which all parties support a given project. When introducing a new assignment, remember that students may have negative feelings toward group work and that past experiences may negatively influence how they perceive the assignment. The same may be said of administrators and other instructors. Regardless, all groups work best when there is broad-based support.

### Keep Team Size in the Range of Three to Five Students

If teams are too small, they may not have the resources they need for complex projects. However, if they are too large, social loafing may become a problem. Research indicates that the most effective collaboration occurs in teams of three to five.

### Make Purpose Central

A team's purpose is central to its ongoing performance and success. You must ensure that purpose stays at the forefront of the team's activities. Occasionally, during the life of a project, ask students to articulate the team's purpose. This activity is a sort of barometer of effective collaboration. If team members articulate the same thing—if they clearly understand and share their purpose—you have a good indicator of how well the team is functioning.

Remember that much of what can be learned in teamwork has to do with teamwork itself, not just producing a document.

### Be Prepared to Work

Dividing students into teams may initially seem like less work (fewer papers to grade); however, instructors may find collaborative work requires a significant time investment up front. Students may need more explanation and clarification as to roles and responsibilities. The group may initially have trouble turning out work, and teams may lack the experience necessary to know how well they are working together or how to judge their work at any given stage. You must be prepared to answer questions and, perhaps most important, to keep generating dialogue among team members.

### Use Metacommunication to Help Prevent Problems

Metacommunication, communication about communication, may serve to prevent misunderstandings among team members. For instance, perhaps not all students are comfortable exchanging phone numbers. In this case, team members could agree ahead of time to use e-mail for project communication. In addition, it is quite likely that students perceive quality communication and reasonable response time differently. Encourage them to discuss such communication issues at the beginning of the project and to establish norms for their own teams.

Modeling professional communication can show students how to work more successfully in a group with busy people, some of whom may have jobs, a family, or both. Metacommunication may prevent misunderstandings between students of different cultural and linguistic backgrounds, too.

### Establish Limits

Almost inevitably, someone will not work well with someone else. Establish ahead of time how you will address student dissatisfaction. There's no doubt that students can benefit from learning how to work with a difficult team member; however, it's not fair (and may be against institutional guidelines) for everyone's grade to suffer as a result of a single person. Deciding ahead of time what the limits are may help facilitate moving a team member to another team. Distribute those guidelines to the teams to help avoid problems in the first place.

### Manage Time

Giving students a deadline helps keep them on task. When getting started with collaborative work, five minutes or so may be enough to get students working together productively while not wasting too much class time. Later, more time may be needed to complete more complex tasks. Consider using brief periods of class time for students to touch base and set goals and agree on time outside class for meeting and performing work.

Although you cannot, and should not, try to control what work gets done and when, making sure group contact occurs during class time will ensure that the group is indeed meeting together at least once per week.

### Establish Multifaceted Assessment

Evaluating student collaborative work can mean a number of different things: evaluating the team's product, or document; evaluating how well the team worked together; evaluating individual performance; and other possibilities. Peer review, as part of the drafting process, or as a wrap-up evaluation of the team's performance, is crucial to both preparing for work outside the classroom and as an opportunity for reflection and deep learning.

### Manage Social Communication

Collaborative work in the classroom means students need to work together; therefore, the instructor must develop ways of letting students do their work while also keeping the class on task. An experienced instructor knows that a certain amount of nonrelevant activity occurs no matter what (i.e., chatting, laughing, someone in one group talking to someone in another, etc.). Just remember that in the workplace such social interaction is not only common but also necessary for building effective team collaboration. Such interactions are building blocks for communication and trust (Paretti and McNair 2008; Thompson and Coovert 2006).

### Take Notes

As with any new assignment, a new collaborative assignment may have unexpected challenges that are only identified once students have begun work. Perhaps the best way to handle those challenges is to make note of them and then make changes as necessary for the following terms. The best collaborations may, in fact, be between teams of students and the instructor, wherein both parties learn.

### Use Realistic Scenarios

Although the assignment itself might not seem like a strategy for class-room management, it is actually critical. If the assignment is reflective of the kinds of work students can expect in the workplace, they tend to work more diligently and ask more probing questions because students view such assignments as authentic and, therefore, credible in their life strategies. If the assignment is not the kind of work they will find after graduation, they tend to lose interest.

## TOOLS FOR COLLABORATION

Software tools in the form of ICTs (information communication tech-nologies) often support collaboration in today's learning contexts. As a general rule, it is best to give student teams a choice of technologies so they can choose the tools that work best for their team members and goals. Help students select tools based on their team needs. To do so, focus on the affordances of tools (that is, the characteristics, both perceived and actual, of technologies that affect how they are used) (Norman 1988). For example, rich communication and lean communication are affordances of tools. Rich media support multiple cues similar to face-to-face communication. The closer a medium is to face-to-face communication, the richer it is. For example, video conferencing is a rich medium as it offers audio and video cues in real time. The more ambiguous or complex a task, the more richness is needed. Lean media support fewer cues (for example, e-mail is a lean medium, as it offers only text with some delay). Lean media can be very effective for communicating concrete information, as they decrease unnecessary cues.

If you would like to provide students with more information on the affordances of technology so they can make informed decisions, we recommend using a taxonomy of ICT affordances developed by Hewett, Robidoux, and Remley (2010):

- Presence awareness is "the degree to which individuals in virtual settings know that others are present or available to communicate."
- Synchronicity is "the length of time it takes for individuals to interact using virtual collaborative technology."
- Hybridity is "the use of tools that combine different elements of communication, such as speech and written language."
- Interactivity is "the extent to which individuals can maintain a dynamic flow of communication across virtual space and interactions made when a tool seems to diminish spatial distance." (Hewett, Robidoux, and Remley 2010, 12)

Have students identify these affordances for each technology they would like to use and require that they make tool choices that, together, provide for each affordance.


## SAMPLE COLLABORATIONS

In this section, we provide you with ideas for collaborative projects in your technical communication class. For each project, we provide suggested setting, timing, skills used, and basic instructions. You should feel free to take these ideas and adapt them for your own class and student population.


*Five-Minute Poster Presentation (One Class)*

*Skills applied through collaboration: audience analysis (public, media, supervisor), choice of media for communication (texting, phone call, e-mail), document design (poster)*

    *Instructor: England, engineering students*

Technical writing classes frequently feature oral presentations, and oral presentations work well as an opportunity for students to learn collaboration skills both inside and outside the classroom.

Give students a scenario relevant to course content. Break students into groups of no more than four or five and provide each group with the means to make a poster. A low-tech, tactile approach England uses is large sheets of butcher paper and markers. Alternatively, students can use PowerPoint and Page Setup to set poster dimensions. Students must meaningfully address the scenario and create a poster that serves as a visual aid.

For example, inform students that they are representatives of the Department of Transportation and that they must identify what they will do upon arrival at an intersection blocked by an accident involving a Department of Transportation dump truck, how to address the media, what to report to their supervisor, and how to report it. Each student must participate, though the group may assign a single person to be the scribe and make the poster. The poster must provide visual aids that help the audience understand the group's message.

### Four-Minute Presentation (One Week)

*Skills applied through collaboration: critical thinking (how their field fits into economic action), audience analysis (communicating with a nontechnical audience), document design (slides)*

*Instructor: England, engineering students*

In the final minutes of class, show students images that represent extraction of natural resources (for instance, coal mining, corn farming, and cattle raising). Instruct teams to identify how civil engineering plays a role in the extraction, refinement, and transportation of that product to market. Suggest that a successful presentation necessitates at least one meeting outside class. Tell students they will have five minutes at the beginning of the next class to fix any last-minute problems. Allow students to use PowerPoint, but limit them to no more than two slides.

This exercise encourages students to think broadly about their chosen profession and how it affects many different human activities. Students must think carefully about what they say because in the course of a week they'll discover quite a bit more than can be covered in four minutes. Students must prioritize and edit in order to fit meaningful information onto only two slides. Students must also think carefully about what graphics to use and about how graphics can replace text, but only in the right context.

### Comparison Memo and Project Proposal
### (Prep Time: Two Weeks to Two Months)

*Skills applied through collaboration: research (potential solutions, costs, scheduling), editing and revising (unified design, voice), managing (schedules, conflicts), technical writing and editing*

   *Instructor: England, engineering students*

Students must write a memo that identifies an engineering issue on campus, then identify at least one other college campus that has addressed the same issue. The students are provided with types of campus engineering problems and choose to address one: flooding, traffic, parking, and so forth. The end of the memo must identify why the comparison makes sense (similar climate, similar number of students, etc.). This memo is an individual assignment.

After returning the graded memos, put students into groups based on what types of problems they identified. All the students who identified parking problems are in a single group, and the same goes for traffic, flooding, and so forth. Groups should not have more than five members, so there is often more than one group per problem type. Instruct each group to identify a single problem from their memos and to then generate a proposal to fix the problem.

Students typically bring information they gathered while working on individual projects. They must decide which information is best and how to organize it. There is too much work for one person, so encourage the students to divide up the work.

### Game Instructions (One to Two Weeks)

*Skills applied through collaboration: audience analysis, writing and designing instructions, usability assessment*

   *Instructor: Brewer, technical communication, business, and engineering students*

Identify a game that is unusual or has been passed down orally—in other words, a simple game that does not have readily available instructions. During class, place students into groups of three to five. Explain to them that information gathering in the workplace is quite often done orally. In other words, if someone in the workplace wants them to write and design instructions, that request, and whatever help is offered, will often be offered orally.

Explain the rules of the game and ask students to play it. The students begin to collaborate in an informal context that prepares them for the more complex collaboration to come. They can ask you questions, as you will be acting in the role of subject-matter expert (SME). After they have played the game, ask them to draft instructions for playing it. Many students quickly produce a paragraph or two of instructions. However, some groups begin to ask questions such as, "Who are we writing this for?"

Cover the theory of writing instructions using their experience thus far for examples of the parts that might be needed, how terminology should be used, and how design and graphics will support usability. Ask students how their draft has fared and then encourage them to edit their instructions. If you wish to extend the principles taught, you can also require that students conduct a usability test by asking representative members of their audience to use the instructions while they observe.

## SAMPLE VIRTUAL-COLLABORATION PROJECTS

Collaboration in the workplace (and even in the classroom) can seldom be classified as either face to face or virtual, as most teams conduct their work with both types of collaboration. Because international collaborations tend to rely more heavily on virtual collaboration, and because most students will work in the global marketplace, we would like to provide you with some models for virtual collaboration as well.

*Infographic and Concept Summary (Five Weeks)*

*Skills applied through collaboration: virtual communication, cross-disciplinary communication, information design for multiple outputs*

*Instructor: Brewer, professional communication, engineering, business, and education students*

Collaborate with faculty members from other disciplines within your university on this virtual team project. Doing so enables you to place students in virtual teams with students they don't already know, with students who are culturally diverse, and/or with students in different disciplines who might approach work tasks differently. You could easily create the same collaboration with only one colleague from another discipline.

Create a project that requires the skills of students from all of the disciplines involved. For example, Brewer collaborated on a project with colleagues from finance and library science. Students researched and created an infographic and a one-page summary on the concept of coinsurance.

Place students into teams of four to six students with at least one team member representing each of the three classes. Students then contribute knowledge to the team based on what they are learning in their respective courses. For example, in Brewer's project, students in the finance course contributed specialized knowledge on the concept of coinsurance.

Students in library science contributed specialized knowledge on resources, and students in professional writing contributed specialized knowledge on the writing and design of infographics and summaries.

Encourage students to choose the technologies that best facilitate their context, such as Skype, Google Drive, Facebook, Dropbox, and so on.

*User Guide (Five Weeks)*

*Skills applied through collaboration: virtual communication, international communication, document design*

  *Instructor: Brewer, technical communication and engineering students*

Collaborate with an instructor at a university or training organization in another country. You can make these contacts through your own networks, visitors to your campus, your international office, and professional organizations.

The purpose of this project is to provide students from different nations and cultures the opportunity to collaborate toward a common goal in globally distributed teams and at the same time create user assistance through simple, clear, and usable documentation.

Place students in international groups of four to five students. Students must share a common language even though they may have varying levels of skill with the language. Task each group with creating a user guide for an online retail site or product. Brewer prefers that early attempts focus on writing and designing the user guide for a novice audience who is not expert in using the selected product. Brewer's students have created guides for sites like Amazon.com and products such as self-composting toilets.

In one such project, Brewer's students collaborated with students at Blue Dots Consultancy in Bangalore, India. Blue Dots' students were in the UTC/GMT +5.5 time zone. Students at Mercer University were in the UTC/GMT –4 time zone. Thus, Blue Dots' students were nine and one-half hours ahead of Mercer students. One of the first challenges for students was to structure a work schedule that could work for all team members.

You can increase the complexity and scope of this project by adding a communication plan, a project proposal, or a document plan.

## ASSESSMENT

Assessment of collaborative assignments must walk the line we mentioned earlier: educational institutions are, by nature, set up to evaluate individual rather than collaborative work. Teachers using collaborative work must be ready to navigate institutional, pedagogical, and individual questions. Furthermore, the assessment will likely include evaluation of the product (the assignment itself) and how well students work together. The latter is important in that workplace collaborations are often assigned based on skill, experience, and how well the team members work together.

In short, assessment may be the portion of collaborative assignments most difficult to navigate and therefore deserves attention ahead of time. An experienced instructor might also leave wiggle room for minor changes in assessment based on challenges that arise as students are working. Consider the following issues in planning your assessment:

- How does collaborative work fit within the established learning outcomes set by your department or institution? Because technical writing courses are taught often by non-tenure-track faculty who may have little administrative support, knowing your school's attitudes toward collaborative work might be the most important step in preparing for this type of assignment.
- How will individual contributions be evaluated, if at all? Or will only the group's product be evaluated? As mentioned earlier, it's common for institutions to be focused on individual grade reporting.
- Will student teams have a leader or manager? If so, will that person be assigned by the instructor or elected by the group? If there is a leader, is that person evaluated the same as everyone else, or differently? For instance, the manager could contribute to the assignment as well as check others' work and establish and enforce deadlines. More work might equal more credit.
- If students are to be evaluated for how well they work together (teamwork requires communication, and technical writing classes are all about communication), who will do that evaluation? Instructors are frequently in the dark about how often and how well individual students contribute work, and other students might not provide reliable (or relevant) data concerning another student. On the other hand, the students working together likely know more about each other than does the instructor. Providing evaluation tools, such as a rubric for evaluating another's performance, ahead of time can alleviate many concerns.

Brewer distributes a performance rubric to her students at the end of collaborative projects (see app. 10.A) in which students are asked to assess both themselves and their teammates. They submit their evaluations via e-mail for confidentiality. It is important that students evaluate themselves as well as their peers, as this encourages clear and realistic reporting of contributions to the project. Place special emphasis on the task lists, as this is evidence of performance, and it is easy to see discrepancies in task lists if they exist.

## CONCLUSION

For an instructor, collaborative work means (initially) more work. There is little doubt that adding a new assignment and methods of assessment will mean more work for you when you may already feel pressed for time and energy. The rewards, however, can be substantial. Collaborative work, constructed well, often results in superior work, learning, and retention. Collaborative work can also mean instructors spend more time asking questions and prodding teams along productive paths and less time grading (grading a few teams rather than many students). Collaborative assignments can also result in learning that simply was not addressed before, or learning that is more readily transferred.

The greatest challenges instructors may face have to do with classroom management or assessment. Many college-level instructors may feel their academic background means they do not have the training or experience to be called *leader* or *manager*. But effective collaborative learning requires leadership. A good leader sets a goal and deliverables and then lets teams develop their own strategies and processes. Teams allowed to work and thrive on their own make mistakes and under the best circumstances learn lessons not otherwise available. Students learn from each other, and although not all that learning is comfortable or "good," it is nonetheless critical to future success. To the extent possible, a good instructor lets teams arrive at the given destination but helps them learn from their experiences along the way. Although the metaphor has become a cliché, we nonetheless know the journey is often more important than the destination.

## DISCUSSION QUESTIONS

1. Which of your course's stated learning outcomes could be addressed with collaborative assignments? Do any of those learning outcomes include concepts that imply collaborative learning? For instance, words like *interdisciplinary* imply working with others.

2. Which of your assignments cannot be done collaboratively? Why?

3. How would collaborative assignments alter, for better or worse, the instructor's work load for your course? What about the first time you use them? What about the second?

4. How would collaborative work contribute to a given learning outcome? We already mentioned the collaboration implied by *interdisciplinary*. What about intercultural, environmental, social, or economic outcomes?

5. What positive learning can be achieved in your course only through collaborative work? For instance, students are not likely to learn much about other people's communication preferences until they work on teams. How will that knowledge help students after graduation?

## APPENDIX 10.A

### Sample Self/Peer Evaluation for Collaborative Work

Section 1. Please circle the answer that best describes your team for each of the three items below:

Did all members of the group share in the team's responsibilities?
- Some members did no work at all
- A few members did most of the work
- The work was generally shared by all members
- Everyone did an equal share of the work

Which of the following best describes the level of conflict at group meetings?
- No conflict, everyone seemed to agree on what to do
- There were disagreements, but they were easily resolved
- Disagreements were resolved with considerable difficulty
- Open warfare: still unresolved

How productive was the group overall?
- Accomplished some but not all of the project's requirements
- Met the project requirements but could have done much better
- Efficiently accomplished goals we set for ourselves
- Went way beyond what we had to do, exceeding even our own goals

Section 2. Describe in detail your specific contributions to the project.

Section 3. Write a brief description of the problems you encountered in working with this group and how they were resolved.

Section 4. Please distribute 100 points among the members of your team, based on each member's contribution to the team's efforts. (Don't forget to include yourself.) Use integers only. No two people should receive the same number of points.

*References*

Bosch-Sijtsema, Petra M., and Anu Sivunen. 2013. "Professional Virtual Worlds Supporting Computer-Mediated Communication, Collaboration, and Learning in Geographically Distributed Contexts." *IEEE Transactions on Professional Communication* 56 (2): 160–75.

Brewer, Pam Estes, Alana Mitchell, Robert Sanders, Paul Wallace, and David D. Wood. 2015. "Teaching and Learning in Cross-Disciplinary Virtual Teams." *IEEE Transactions on Professional Communication* 58 (2): 208–29.

Burnett, Rebecca E., L. Andrew Cooper, and Candice A. Welhausen. 2013. "What Do Technical Communicators Need to Know about Collaboration?" In *Solving Problems in Technical Communication*, edited by Johndan Johnson-Eilola and Stuart A. Selber, 454–78. Chicago, IL: University of Chicago Press.

Davis, Marjorie. 2015. Re: Collaboration. Retrieved from CPTSC E-List. June 15, 2015.

Eison, James. 2002. "Teaching Strategies for the Twenty-First Century." In *Field Guide to Academic Leadership*, edited by Robert M. Diamond and Bronwyn Adam, 157–73. San Francisco, CA: John Wiley & Sons.

Hewett, Beth L., Charlotte Robidoux, and David Remley. 2010. "Principles for Exploring Virtual Collaborative Writing." In *Virtual Collaborative Writing in the Workplace: Computer-Mediated Communication Technologies and Processes*, edited by Beth L. Hewett and Charlotte Robidoux, 1–27. Hershey, PA: Information Science Reference.

Meloncon, Lisa, and Peter S. England. 2011. "The Current Status of Contingent Faculty in Technical and Professional Communication." *College English* 73 (4): 396–408.

Mudge, Bob. 2015. "Why Collaboration Is Crucial to Success." *Fast Company*, July 20.

Nidumolu, Ram, Jib Ellison, John Whalen, and Erin Billman. 2014. "The Collaboration Imperative." *Harvard Business Review* 92 (4).

Norman, Donald A. 1988. *The Psychology of Everyday Things*. New York: Basic Books.

Paretti, Marie C., and Lisa D. McNair. 2008. "Communication in Global Virtual Activity Systems." In *Handbook of Research on Virtual Workplaces and the New Nature of Business*, edited by Pavel Zemliansky and Kirk St.Amant, 24–38. Hershey, PA: Information Science Reference.

Thompson, Lori F., and Michael D. Coovert. 2006. "Understanding and Developing Virtual Computer-Supported Teams." In *Creating High-Tech Teams*, edited by Clint Bowers, Eduardo Salas and Florian Jentsch, 213–41. Washington, DC: American Psychological Association.

Vygotsky, Lev S. 1978. *Mind in Society: The Development of Higher Psychological Processes*. Cambridge, MA: Harvard University Press.

# 11

# TEACHING USABILITY TESTING
## Coding Usability Testing Data

Tharon W. Howard

## INTRODUCTION

Since the late 80s and early 90s, usability testing has grown from a fringe activity in which technical communicators might occasionally partici-pate to the point at which it has become a mainstream skill required for all professionals. Beginning in 2010, if a technical communicator wished to receive certification from the Society for Technical Communication and be recognized as a professional in the field, proficiency in usability testing was one of the six competencies candidates had to demon-strate (see Society for Technical Communication 2013: "Certification Maintenance Policies" at http://www.stc.org/images/stories/pdf/main tainingcertification.pdf and http://www.stc.org/education/certifica tion/certification-main). Furthermore, in their 2013 survey of 185 tech-nical and professional communication programs in the United States, Lisa Meloncon and Sally Henschel found a substantial increase in the number of programs requiring courses in usability testing as a gradua-tion requirement, compared to Harner and Rich's 2005 survey of TC programs (Meloncon and Henschel 2013).

To adequately prepare students for industry practices and work-place expectations, instructors in today's technical communication classrooms must focus explicitly on integrating user-centered design processes and usability testing into their assignments. Indeed, in their textbook, *Technical Communication*, Laura Gurak and John Lannon go so far as to define the whole of technical communication as "user-centered communication" (Gurak and Lannon 2010, 5). But if you're new to teaching technical writing or if you would like to make usability test-ing a more integral part of your class—yet you've never had a course in usability testing—what do you do? This chapter focuses explicitly on helping instructors learn how to integrate usability testing into their document-development processes. Specifically, the chapter explains the

DOI: 10.7330/9781607326809.c011

basic phases instructors must go through in assignments such as writing instruction manuals and producing documentation sets so these assignments can be integrated into a course's curriculum. Also, since experience has shown that the most difficult phase of the process for both students and new instructors is learning to code the data collected in ways that lead to *actionable* recommendation reports, particular attention is given to coding and analyzing the data.

## GETTING STARTED

Fortunately, for those new to usability testing, there are a large number of resources available that "make the rocket surgery easy" (Krug 2010), and there are more resources that you can use to learn more about usability testing research methodologies than it's possible to deliver in a single book chapter. And for the more experienced instructor, there are a number of books I keep on my shelf as "go-to" resources when I need examples to show students and when I want to assign short readings to help raise students' awareness of usability as a concept. One of my favorite techniques when we are getting ready to start the unit on usability testing in my technical writing class is to have students read one or two stories from Steven Casey's (1998) *Set Phasers on Stun. Set Phasers* and Casey's (2006) similar anthology, *The Atomic Café*, are collections of short, pithy, and shockingly true stories about situations in industry in which the usability of a product or service has gone horribly wrong. For example, "Set Phasers on Stun" describes a poorly designed software interface for a chemotherapy machine used in the treatment of cancers. Due to its poor usability design, the system's interface not only allows medical technicians to deliver lethal doses of radiation to patients, it actually encourages them to do so. Since most of the students in my technical writing courses are preparing for careers as engineers or scientists, I ask them to read a couple of stories from Casey's books, and then we have a short, in-class discussion about how the usability testing studies we will be doing in the course have significant practical implications for the design and engineering work the students will be doing after graduation. It's a great motivator for STEM students who don't always see why they need a humanities-based course in technical writing.

As far as learning the methods of usability research goes, I have three texts I always recommend. First, even though it was originally published in 1993, many in the field still consider Joe Dumas and Ginny Redish's *A Practical Guide to Usability Testing* to be the "bible" for usability practitioners (Dumas and Redish [1993] 1999). Republished in 1999, it is the "golden oldie" that continues to provide technical communicators with a no-nonsense, practical introduction to how to conduct a usability study. The other two books I recommend for research methods and for examples of the parts that make up a usability study are Jeff Rubin and Dana Chisnell's *Handbook of Usability Testing* (Rubin and Chisnell 2008) and Carol Barnum's (2011) *Usability Testing Essentials.* I've used excerpts from all three of these in my technical writing classes since the early 90s, and my undergraduates have never experienced any difficulty understanding the material. However, if you're teaching at the graduate level or if you are looking for a text that delves more deeply into the quantitative side of usability research methods, I'd recommend Tom Tullis and Bill Albert's *Measuring the User Experience* (Tullis and Albert 2013). Tullis and Albert provide a useful introduction to metrics that can be used in order to create quantitative measures of users' experiences with a product or service that can then be statistically analyzed and validated. These are skills I certainly teach in my graduate seminar on usability testing, and the book is worth mentioning as a useful resource for instructors who want to take a deeper dive into usability research. However, since my undergraduate technical writing classes are more focused on qualitative research designs, we rarely use probability or correlational statistics and stick mainly with measures of central tendency (e.g., mean, mode, standard deviation, etc.).

In addition to book-length resources, I would be remiss if I didn't mention a few of the online resources I frequently use. Printed materials are great, but since usability testing essentially involves asking a representative sample of users to "think aloud" as they perform a typical task with a product (such as an instruction manual), the work students will be doing mainly involves visual observation of the users' behaviors. Consequently, it's extremely helpful to have a collection of video clips of users conducting actual think-aloud protocols that you can show students. Personally, I'm able to use videos of actual research studies I've done for industry clients, and as you conduct more and more studies with your technical writing students, it's a good idea to begin building your own personal library of video examples (just be sure to have the participants' informed consent statements and video permission forms on file so you can legally show them in your classes).

Yet, when you're first starting out, you're obviously not going to have clips to show students, so you'll probably need to depend on YouTube and other online resources. Sadly, because of the federal laws governing how institutions can use the research data they share with the public and because few companies are willing to release videos of people using early (and thus flawed) prototypes of products they are building, there simply aren't many examples of think-aloud protocols on the Internet. There are, however, five examples I found that could be used to generate class discussions about how to conduct think-aloud protocols until you're able to build your own personal library:

- *Rocket Surgery Made Easy by Steve Krug: Usability Demo* by Peachpit TV at https://youtu.be/QckIzHC99Xc
- *Usability Testing of Fruit* by Blink UX at https://youtu.be/3Qg80qTfzgU
- *Example Usability Test with a Paper Prototype* by BlueDuckLabs at https://youtu.be/9wQkLthhHKA
- *Dumas, Loring: Moderating Usability Tests* by Elsevier at http://booksite .elsevier.com/9780123739339/?ISBN=9780123739339/
- *The Moderator's Survival Guide* by Donna Tedesco and Fiona Tranquada at http://www.modsurvivalguide.org/videos/

The other online resources I recommend are

- Usability.gov at http://www.usability.gov/
- Usability Net at http://www.usabilitynet.org/home.htm
- The UXPA's UX Resources page at https://uxpa.org/resources /external-websites-organizations

All three of these websites are one-stop shops for a broad range of topics involving usability testing. All of them give excellent introductions to how to conduct different types of usability tests, they all have templates and examples students can use, and all of them are keyword searchable.

### WHY TEACH USABILITY?

So far, I've made the argument that usability testing should be incorporated into technical writing classes because it's considered mainstream practice in industry today. Educators have a responsibility to prepare students for the skills desired by modern companies, and there are many reasons industries are seeking a workforce educated in usability testing. Obviously, usability testing the information products companies produce reduces the errors found in those products. This, in turn, helps companies reduce their return rates on products and avoids costly restocking fees. It also cuts costs by reducing the volume of customer-support calls users make because they require assistance with installation, maintenance, troubleshooting, or other functionality the product is supposed to provide. Improving the usability of information products also enhances customer satisfaction, which leads to increased brand loyalty, word-of-mouth marketing, and ultimately enhanced sales. In fact, as Randolph Bias and Deborah Mayhew have shown in their anthologies on *Cost-Justifying Usability* (Bias and Mayhew 1994, 2005), these are just a few of the ways industries can easily justify why they find graduates with usability testing experience desirable.

Yet, while I'm compelled by the desire to make students more marketable, that goal alone wasn't and still isn't a sufficient condition for me to incorporate usability testing into my classes. As Caroline Miller has also argued in "What's Practical about Technical Writing," simply replicating industry practices in our classrooms doesn't mean we're not encouraging industry's "malpractices" (Miller 1989, 17). Beyond the industry rationale, there must be strong pedagogical reasons for making usability testing processes a part of our technical writing instruction. The main reason I incorporated usability testing into my technical writing classes is because I simply couldn't find a more effective means of teaching students audience awareness and the impact audience awareness has on a document's style, word choice, organizational structure, and page design.

As writing teachers, we incorporate all sorts of pedagogical techniques into our classes intended to help students develop a sense of audience awareness. We use invention heuristics from classical rhetoric as part of the writing processes we teach students to force them to think about and analyze the needs of their audiences. We ask students to read and critique both good and bad examples of the types of documents they are producing in order to help them develop an understanding of how organizational structures and the visual designs of pages help readers (and users) navigate through texts. And we valiantly scribble sentence-level feedback in the margins of the drafts and final documents student writers produce, desperately hoping our "voices" in that marginalia will help the students develop an understanding of how their syntax and stylistic choices produce a response in their audiences.

These time-honored methods play a critical role helping students develop a sense of audience awareness in our classrooms and surely have an effect. Yet we also know more can be done. We know from studies such as Mike Rose's (2009) work on writer's block that we have little control over the ways students interpret what we scribble in the margins of their papers. And what experienced teacher hasn't come to suspect that students have grown a bit "teacher deaf" by the time they reach us in our technical writing classrooms? How many of us bring in industry professionals to offer testimonials to students—testimonials that support the lectures we've been giving students about the importance of audience—because we aren't entirely sure students find our voices credible?

Requiring students to usability test their own documents cuts through the credibility gap problem. More important, it puts students as technical writers in the closest possible proximity to their audience in a way that has the greatest potential for impact. As an instructor, I can write *awk* or *unclear* in the margins of a student's paper. I can even write "This will confuse your users," but I also know students have had years of training in educational settings, which has taught them how to emotionally distance themselves from the psychological impact a teacher's comments in the margins can have on them. Yet put that same student in a situation in which she's written a real manual, intended for real users, and then ask her to watch a video in which she observes those users becoming frustrated and annoyed with HER lack of clarity, HER wrenched syntax, or HER failure to provide adequate navigational cues on the page . . . confront students with real users' voices instead of teachers' by asking them to user test their documents and the exigencies of audience awareness in technical writing can have a new relevance for students.

## LOCATING ASSIGNMENTS AND THE FIVE PHASES OF A STUDY

In order to begin integrating usability testing into your course, you'll first need to select a writing assignment that results in a product intended for use by actual users. Since nearly all technical writing courses involve writing instructions and documentation sets, the user manual is probably the best and most obvious choice. I've also used this assignment with recommendation reports and proposals in classes when I've had students working with real clients who were the audiences for these documents. However, because the logistics of working with actual clients in a service-learning environment is already a complicated and complex undertaking, I recommend using an assignment for which students produce an instruction manual as a starting point.

Next, when choosing the products or services for which the students will be producing their documentation, it's a good idea to consider how much access the students will have to users who can participate in their usability studies. In other words, if you ask nineteen undergraduate students to write manuals for some obscure process that is only used by three or four people, students won't be able to find enough volunteers to serve as participants in their studies. Consequently, find processes and procedures for the students to document that are available to their roommates, friends, and family. For example, you might require that students produce manuals on topics like how to create a Pandora account, create stations, and add variety to the stations. Or students could produce manuals on how to use the Citation Machine at www.cita tionmachine.net to create APA and MLA citations for research papers. These services are available without cost on the Internet, so there are no access barriers or costly licensing issues for students. Also, it's entirely likely that students will have no difficulty finding other students at your school who could be convinced to volunteer to user test students' manuals. I've also had success finding excellent documentation projects by working with the librarians or the instructional-technology centers at my own institution. They often offer publicly available and widely used services in the library and computer laboratories that could benefit from documentation. Whatever topics you choose, however, the main criterion to keep in mind is that access to users is a *sine qua non* for the project. Students can't conduct any usability tests without easily obtaining access to volunteers for their studies.

Once you've settled on the products or services students will document, it's necessary to break the process of conducting a test into its specific phases. A common misconception regarding usability testing is that students can just show their manuals to their friends, get feedback, and then claim they've user tested their manuals. That makes about as much sense as putting lipstick on a pig. Without following a rigorous research methodology that ensures the students and their users have been exposed to and have thoroughly examined all the potential usability problems with the manuals, just getting feedback from users can't legitimately be called *usability testing*.

During the class in which we begin the usability testing assignment, I break usability testing down into five phases: (1) establishing the research questions that must be addressed for the product being examined, (2) planning the test so it collects reliable data from representative users in a valid context, (3) collecting the data, (4) analyzing and coding the data, and (5) generating a recommendation report that focuses on actionable results. During this same class period, I also introduce students to the ISO 9241-11 definition of *usability*: "The extent to which a product can be used by *specified users* to achieve *specified goals* with effectiveness, efficiency and satisfaction in a specified context of use" (Usability.net, emphasis added).

When I introduce this international standard to the students, I emphasize "specified users" and "specified goals" to the students for two reasons. First, I want to impress on them the importance of audience and the fact that there are inappropriate audiences (users) for manuals. This sets up our discussion of representative sampling methods, which will come later. Second, and more importantly, I focus on the "specified goals" part of the definition by asking the students to think about who gets to set the goals for the manual's use. Do the authors decide or do the *users* decide what will be the goals for how a manual is used? This is often a question students have never considered before; they usually assume they, as authors, have control over their manuals. However, it doesn't take much discussion before they quickly realize that they, as authors, don't have absolute control over users. As authors, they can encourage users to adopt certain goals for the manual through the language they use, the titles they use for sections of the manual, and so on. They can "invoke" (Ede and Lunsford 1984) the roles users play through techniques such as gamification, for example. But ultimately, it's the users who set their own usage goals, and manuals that fail to either accommodate those goals or fail to encourage users to adopt alternative goals are unlikely to be judged usable by the audience. Thus, once again, my pedagogical goal here is to help raise students' consciousness of the critical role audience awareness plays in their success as technical writers.

In this same class period, I also introduce students to the concept of the think-aloud protocol. They learn that asking users to perform a think-aloud protocol with their manuals does *not* mean asking them to read each sentence of the manual aloud and then comment on the sentences. Instead, they learn think-aloud protocols essentially involve asking a *representative* sample of users to think aloud as they perform a *typical* task with the manuals.

I invite the students to put on their "white lab coats" and to think like a researcher. We discuss what it means to have a representative sample. How might sampling bias the study or undermine its reliability and validity? Will people who may have some experience with the product the manual covers be typical users? If so, how might their use of the manual differ from the novice user who has never experienced the product before? What parts of the manual might more experienced users read compared to novices? We also discuss the ways changing the types of tasks we ask users to perform with the manual might influence or bias their use of the manual. For example, is there a difference between the ways users might go through the manual when they are installing the product for the first time compared to the ways they might use it if they're simply updating a feature in the product? It's also during this class that we watch sample videos of users' think-aloud protocols like the ones listed in the "Getting Started" section of this chapter in order to help the students better understand the observational nature of the data they will be collecting during their studies.

*Establish Research Questions*

Asking students to collect information from users without ensuring they have an extremely clear understanding of what types of data they need to collect is a recipe for disaster. Without clear research questions to drive their studies, students' user testing will lack the rigor necessary to prevent their work from devolving into a series of merely anecdotal observations.

Setting clear research questions with the class is also an important pedagogical opportunity because it allows the instructor to review many of the principles of effective manual design students have hopefully already learned by creating their instruction manuals. Consequently, during this portion of the class, the students and I discuss what they, as the authors of the manuals, want to know in order to improve their grades on their manuals. I give the students credit for completing a version of the manual that can be usability tested, but in order to keep them motivated to do the testing, I don't give them a final grade for their manuals until they have completed the revisions necessitated by the usability studies.

Each class can generate slightly different research questions depending on the type of product or service students are using for their manuals. However, the research questions the students and I generate roughly fall into the following four categories:

- **Effectiveness of page layouts and navigation**: Can users find what they need easily? Can they scan the page, locate specific information zones on it, and then go directly to the information they are seeking? Are the headers clear to users and appropriately labeled? Are users able to leave the page, complete a procedure described in the manual, and then relocate their place on the page?

- **Usefulness of information**: Does the manual provide the content users want? Is the content clear, or does the user have difficulty interpreting it? Is the information accurate, and are the users able to follow the procedures correctly? Is there a good balance of visuals such as screen captures and text?

- **Effectiveness of the style**: Is the information "chunked" into manageable sizes, or do the users complain that the text is "wordy" or "text heavy?" Are there specific sentences or paragraphs that distract or confuse the user? Are the fonts, use of color, or use of graphics/visuals distracting or problematic? Are the conventions in the manual, such as typing optional information inside brackets (as in "[type your text here]"), alerts, and warnings, clear to users?

- **Successful completion criteria (SCC) rate**: Are the users able to accomplish the tasks they are supposed to accomplish? Do they struggle to complete tasks, and if so, how much? How do they feel about their experience as they work on the tasks?

The first three categories are mainly review material and involve traditional principles involved in constructing documentation sets, starting at a high level of abstraction (i.e., overall page design) and then moving down to sentence-level issues by the third category. However, the fourth category here is intended to help students begin developing the quantitative metrics they will be using in their usability studies. By asking students to define the criteria necessary for the manual to be considered a success, and then by introducing the concept of a rate, which will be used to score how well the manual meets the criteria, my goal here is to disabuse students of the notion that they are merely collecting anecdotal reading responses from users.

Once the students and I have discussed all the key tasks their users should be able to complete with their manuals, I introduce them to the traditional four-point usability severity score used by industry practitioners to rate how well users were able to complete tasks. There are a number of variations on this scale (see for example Nielsen 1995 or Sauro 2013), but the one I use with technical communication students on their manuals is as follows:

1.  User completed task without difficulty.

2.  User experienced minor problems completing the task due to cosmetic or similar issues. Users were slightly annoyed, but not frustrated.

3.  User experienced major, significant problems completing the task. The user was frustrated but was able to ultimately solve the problem.

4.  User experienced a catastrophic problem and could not complete the task.

Although, technically speaking, the task-severity scale is part of the data-collection instruments students don't have to begin developing until the planning phase, I think it's a good idea to introduce it here during the research-questions phase. It's important for students to be able to connect the dots between the research questions they're asking and the data they're collecting. Introducing the severity-rating scale at this point makes that connection more explicit. Plus it introduces an element of methodological rigor to the process that helps ensure they understand they are conducting an actual empirical research study.

### Plan the Study

Planning to actually execute the usability study requires four distinct parts. Students must complete the following tasks:

- define participant profiles and recruit participants
- create data-collection instruments
- produce informed consent and video release forms
- write instructions for users, scenarios, and tasks

### Participant Profiles and Recruiting

Because we've already discussed the idea of representative users and the impact different types of users can have on the way the manuals are used, students are already prepared to create their user profiles. The profiles are short documents in which the students describe their users and provide statements of the criteria an individual must meet in order to participate in the research study.

Also, it's at this point in the class that I introduce students to the discussion of how many participants they need to have in the study. Since students are often used to survey research with large numbers of subjects in their population, I take students to the Nielsen/Norman Group website at http://www.nngroup.com/articles/why-you-only-need-to-test-with-5-users/ and show them the visual there. On this site, Jakob Nielsen (2000) explains how, mathematically, usability studies don't need more than five participants in a study in order to capture 80 percent of the major usability problems in a product. As long as the participants in a study are *representative*, adding more users to the study isn't going to allow a researcher to observe many new behaviors.

### Data-Collection Instruments

We've already discussed one of the data-collection instruments students will employ when they use the four-point severity scores as participants complete tasks using the manual. There are two additional data-collection instruments students will need to develop: the prestudy questionnaire and the poststudy questionnaire. For the prestudy questionnaire, students must develop the questions they will be asking participants about their backgrounds. These questions should collect basic demographic information such as age, education level, sex, and so on. Also, the questions should collect data on the users' previous experiences with and their attitudes toward the product or service documented by the manual.

The poststudy questionnaire should include questions users will be asked after they have used the manual to complete their tasks. One set of these should be Likert-scale questions that ask users to rate the manual's ease of use for each task they completed using the six-point scale below:

Very Easy    Easy    Somewhat Easy    Somewhat Difficult    Difficult    Very Difficult

During the analysis phase, students convert the participant's "Very Easy" response into a 6, "Easy" into a 5, and so on. This provides them with more quantitative data they can use to assess a manual's ease of use. And of course after they ask participants to rate the ease of use for each task, they will also ask the users to explain why they gave the rating they did. Often, it's these explanations that produce the most quotable moments in a study, so I always advise students to record their poststudy interviews so they can capture the participants' comments for their recommendation reports. In the poststudy questionnaire, students also collect data from users about their overall experiences. For example, how did they feel about the navigation system the manual used? Did they have difficulty finding the information they were seeking, and if so, how would they recommend revising the manual?

### Informed Consent Statement and Video Release

Typically, students in a class don't need Institutional Review Board approval to conduct a usability study in a class project (unless they plan to publish or present their findings), but instructors should always check with their local IRBs just in case. Nevertheless, ethically it's essential to provide participants in any usability study with information about how their data will be used, what risks may be involved by their participation, how their privacy and identity will be protected, and for what purposes the study is being conducted. Also, before video or audio recording participants, it's important to secure each individual's permission. Since I require that my technical writing students record the participants in their studies, the students must provide the participants with both an informed consent statement and video release form. The students must submit copies of the signed forms to me when they turn in their reports. Appendixes 11.A and 11.B provide examples of the templates students edit for use in their studies.

### Instructions, Scenarios, and Tasks

Because students are conducting an empirical research study, it's important that all the participants in the study receive the same instructions. As a result, the students must develop written scripts and handouts to use with participants during the studies. After the students have gone over the informed consent statement and video release forms with the participants, they then ask the participants to answer the questions on the prestudy questionnaire. At that point, participants are ready to receive the instructions, which marks the beginning of the protocol-analysis phase.

Because participants are usually nervous and unsure about what is going to be happening, one of the critical goals of the instructions must be to put the participants at ease so they will feel comfortable expressing their honest opinions. No one likes to be tested. So one of the major objectives of the instructions is to help the participants understand *they* aren't being tested; the manual is the object of study, and it is what is being tested. Below are sample instructions to be read at beginning of the think-aloud protocol

### Preliminary Instructions for Student/User:

[ *To be read by the study administrator to the user*]

First, let me begin by thanking you once again for volunteering to help me with this project. I am in the process of studying how people use instruction manuals. The sample manual we're going to be using today is a prototype, meaning it is a draft that is in the early stages of development. With your help, I hope to offer suggestions to people who wrote the manual to help them improve it.

Do you have any questions?

It is imperative that you understand I am **NOT** testing you. In fact, you are aiding me in studying the manual. Even though I will be asking you to complete a few simple tasks, there are no right or wrong answers in this study. Because I am trying to understand what works well and what doesn't work well in manuals, it is important that you voice your thoughts as you try to perform these tasks. Harsh criticism will not offend me; in fact, it can offer a great deal of information about your experience. On the other hand, I am also extremely interested in what you like about this text. Basically, I just want your opinions.

All I request is that you say out loud what you are thinking and feeling as you are using the texts. It's also very important that you use the manual the way you would normally. If you would normally skip pages, then please do so. Please don't feel like you have to read everything just because I'm here. Try to do what you would normally do.

As you use the manual, I may occasionally interrupt to ask questions about what you are doing and thinking.

Do you have any questions?

Another key point I try to make with students here is that they should try to never refer to the usability study as a *test* when working with participants and that they should avoid referring to the study participants as *subjects*. In fact, it's for this reason we don't refer to the prestudy questionnaire and the poststudy interview as the more traditional *pretest* and *posttest*. It's already difficult to make participants feel comfortable enough to be forthright and honest during a protocol analysis, so using language that suggests to them they are "guinea-pig" subjects in a test only exacerbates the problem.

In addition to instructions for the protocols, usability study participants need scenarios. Scenarios are short stories that provide users with realistic, engaging information they need in order to complete the tasks students ask them to perform. Ideally, your scenarios for the task can be really short, sometimes only a sentence or two. For example, if you were conducting a usability test of the Hotels.com website, you might give a scenario like, "Imagine you need to go to Charleston, South Carolina, for a business trip, and you're going to be staying the nights of February 15–17." In this case, users need to know the name of the city where they will be staying and what nights they will be there in order to use the Hotels.com website, and the scenario provides the information to them in a realistic fashion.

However, sometimes the scenarios can become quite complex. For example, one of my classes was working with a real client, the Office of Sponsored Programs (OSP), at the University. OSP is the organization that handles grant proposals from faculty to government agencies (like NSF), large foundations, and corporations. As part of the OSP workflow, there is a form known as the Proposal Processing Form, which faculty must fill out before their proposal can be submitted to an external organization. Unfortunately, the Proposal Processing Form is extremely complex, full of technical jargon faculty don't understand, and difficult for faculty to complete. So the class took on OSP as a client, and the students' job was to help make the Proposal Processing Form easier for faculty to use. Below is the scenario for that usability study.

*Scenario for Proposal Processing Form Study*

Imagine that 2 months ago, you received a request for proposals from the Google Foundation which provides funding to help you purchase computer software, travel to attend workshops, and graduate student support aimed at improving the delivery of content in your courses. The purpose of this grant is to encourage faculty to incorporate their personal research into their classroom pedagogies. Your chair required all faculty in your department to respond by submitting a proposal.

You wrote a proposal entitled "Multimodal Courseware Development" in collaboration with a faculty member from the Pearce Center for Professional Communication named Trisha Flounder (her dept. number is 0523, the Center number is 0775, her username is trishf, and employee id number is 001234). You expect to share equally in the amount of time you invest in the project. Your proposal is basically to develop digital videos which you will use in your classes; you'll provide subject matter expertise and Trisha will train and supervise the MA graduate student who will actually produce the videos; in other words, you're providing 75% of the subject matter expertise and Trisha is provided the remaining digital instruction experience.

In your proposal, you request $500 to purchase video editing software. You also request $180 to pay for 3 people to travel to Columbia, SC to attend a training workshop. You also request $5,000 for a graduate student stipend for 1 semester's worth of support, plus 0.9% in fringe. Finally, you request $3,000 in salary for yourself, plus 28.5% fringe. Note that the funder's request for proposals says that "cost sharing is encouraged but not required." In your proposal, you state that the project will start on July 1, 2010 and end Dec. 31.

Much of the information provided in the scenario above probably doesn't make a lot of sense outside the context of creating an application for a grant proposal. But the point here is that because the Proposal Processing Form required so much detailed information in order to complete it correctly, and because participants in the study didn't come into the study having already written a proposal they wished to submit to OSP, this particular scenario needed to be quite elaborate.

Ideally, the scenarios will have the tasks users need to perform embedded in them, as was the case with the Proposal Processing Form above. All we needed to do in order to begin the task in this case was to provide users with the form and ask them to complete it. Unfortunately, with manuals that document a number of different types of procedures, it's often the case that you'll need to need to come up with additional scripts that ask users to complete a different task for each procedure. For example, I once had to test a website that was designed to help psychology faculty teach their courses. The site had different procedures for logging into the site, adding readings and content to the course, creating online quizzes, checking students' grades in the class, and so on. In situations like this, it's not possible for a single scenario to drive all the tasks. Students may need to extend their initial scenarios to cover additional tasks. For example, in the case of the psychology website, I began with a scenario that told faculty they were creating a syllabus for a course that would be taught on the new website. Then for each additional task we needed the users to perform, I had to create a new set of instructions like the following:

For this portion of the study, please assume that you're still working on your class syllabus, and you decided that you need to assign a quiz for Chapter 1. You know from the sales rep. for the website that there is a way you can have the website provide randomly selected questions from a set of pre-existing questions, so you decide to try to use it.

> Please create and assign a quiz called "Chapter 1 Quiz" for Jan. 20. Please use 3 questions selected randomly from the Chapter 1 test bank.

Once again, because it's important that all the participants in the study receive the same instructions, you'll need to spend time in class helping students think through all the scenarios and tasks they will need to give their participants.

### Collect Data

Once you've finished with the plan-the-study phase, students should have everything they need in order to actually conduct their studies and can begin collecting data. They should have recruited five participants who meet their user-profile criteria, they should have a good idea of the type of information they need to collect in order to answer their research questions, and they should have all the handouts, questionnaires, and test scripts they need in order to administer the study. As the instructor, my role during this phase is to help them collect the data—that is, to share tips with them on how they can collect the data so it can be easily analyzed and quickly incorporated into their recommendation reports.

One of the difficulties experience has shown students will face during the data-collection phase is keeping participants talking during the protocol analysis. In the first place, participants don't always understand what they're supposed to do when they are asked to think aloud while they're performing a task. So one of the techniques I share with students is to model the kind of behavior they want participants to engage in during the protocol. In other words, I advise them to bring a textbook with them to the study and give the participant a thirty-second demonstration of how they would use the textbook, thinking aloud and giving their opinions on different page-design elements and passages in the text. In this way, they can signal to their participants the kinds of information they hope to collect from the participant.

Next, we also watch digital videos of me administering protocol analyses during research projects. As we watch these videos, I pause the videos in order to illustrate the kinds of techniques I'm using to elicit information from participants. For example, participants who get stuck with a task and have difficulty performing it often stop talking. After a three- or four-second pause, I prompt the user by asking a question like "What are you thinking?" or "How are you feeling right now?" I try to impress upon the students how important it is for them to keep the participants talking because if they don't know what their users are thinking, they won't have any data to use to write their recommendation reports. Thus, while we're watching the videos, I try to illustrate how to encourage users to continue to vocalize their thoughts without biasing them.

Another issue I try to illustrate by showing the video clips in class is what to do when a user experiences a catastrophic error and becomes so horribly lost in a process that they are unable to complete the task. Students' inclination is to want to intervene and help the user solve the problem. However, doing so jeopardizes the potential collection of valuable data that could lead to an innovative recommendation on how the manual might be revised. So in order to help the students learn how to negotiate this situation, I show them videos in which I tell participants, "Please do what you would normally do and thank you for trying this manual. I realize you're frustrated, but this is exactly the kind of problem we were hoping to uncover." As a class, we also set a maximum time limit on how long we will allow a participant to struggle before terminating the task and moving on to the next one. Usually I advise students to set a two-minute maximum time limit, although this can be increased significantly depending on the complexity of the task.

How to take notes during a protocol analysis and collect observational data is another topic we discuss during class. I explain to the students that the goal of the recommendation report will be to present the reader with highlights from the usability study that will compel the writers to make changes in their manual. They should highlight as many of the most pithy, insightful, or constructive user comments as possible. And they should do so in a way that makes it easy for them to go back to their video recordings and collect those clips. To help them take notes during a study, I instruct students to create a note-taking page divided into three columns. In the first column, they write the time code from their recorder in which a comment occurred. With video cameras, this code is usually indicated by hours, minutes, and seconds of the video. The students must write down when the comment occurred so they can easily go back and find it later. In the second column of their note-taking page, they write down a one-letter code that indicates the type of comment they are recording. For example, if a participant is making a recommendation about how the manual could be improved, the observer would write an $R$ in the column to indicate this type of comment is a recommendation. Other codes include $S$ for start task, $E$ for the end of a task, $P$ for a positive comment, $N$ for a negative comment, $V$ for a comment about navigation issues, and so on.

## SAMPLE CODES USED TO TAG COMMENTS DURING A PROTOCOL

S = Start task

E = End task

N = Negative comment from user

P = Positive comment from user

R = Recommendation for improvement from user

H = User needs help

X = User gives up, abandons task, or has catastrophic error

I = Insightful remark/metacomment by user

O = Logger makes observation

Q = Question asked by administrator

A = Administrator stops task/intervenes

U = Administrator gives ease-of-use scale at end of task

M = Miscellaneous event/comment

D = Administrator gives directions on test (not help on task)

The third column, which is the widest on the page, is used for describing the type of comment the user made and other observational data. The researcher might write, for example, "User couldn't find the term 'getting started' in the table of contents." The third column is also where the researcher records the task-severity score. When a task ends, the researcher is supposed to give it one of the four-point task-severity scores, and they record it in that third column; that is, they give it a score of 1 if the user completed the task without incident or 4 if the user was unable to complete the task and experienced a catastrophic failure.

During the class on data-collection techniques, I also try to help students with collecting data during their pre- and poststudy interviews. Frequently, students assume that since these are in the form of written questionnaires, they should just hand the paper documents to the users and ask the users to fill out the information. However, this approach is not only boring and difficult for the participant, but it denies the researcher the opportunity to collect richer data from the participant. If the researcher uses the questionnaire sheet as a note-taking device and interviews the participant, writing down the participant's responses on the sheet, they are much more likely to get more thorough and richer data from their participants. Plus, the researcher has the opportunity to clarify what they're asking when participants are confused by the question. Or, conversely, they can ask participants to clarify their answers if the researcher doesn't understand.

## Analyze and Code

Coding and analyzing the data is probably the most overlooked and seldom-taught aspect of usability testing methodologies. Yet a well-done analysis of qualitative data can do more to enhance the rigor of research than any of the other four phases of the process. Teaching students to code qualitative data is what will allow them to get the very most out of their usability studies. Sadly, however, I fear that many teachers and even usability researchers don't methodically code the protocol data they collect. They simply go through their protocols and highlight the most interesting comments users make. They just capture the most opinionated remarks that occur during a protocol and ignore the rest. I can't deny there is some value in this approach, and it will reveal some potential issues with a manual that must be addressed. However, this "cherry-picking" approach to analyzing qualitative data is subjective and lacks methodological rigor. When our technical communication students are learning to analyze qualitative data (many for the first time), they must learn appropriate techniques. As Keith Grant-Davie has observed,

> Coding organizes data, allowing researchers to abstract patterns by comparing the relative placement and frequency of categories. It gives them a system by which to demonstrate these patterns to other readers, and it provides researchers with a perspective from which to view the data so that the coding can directly address their research questions. (Grant-Davie 1992, 272–73)

In other words, failing to teach our technical communication students how to code denies them a means of finding patterns in qualitative data and steals their ability to communicate those patterns to their audience.

Usability testing professionals utilize many different types of coding strategies, but since most of these require more time and resources than are available for undergraduate students in a technical writing class, I teach students the classic "open" and "axial" coding approach developed by Anselm Strauss (1987) in *Qualitative Analysis for Social Scientists*. The coding technique for analyzing the qualitative data collected during the think-aloud protocol basically begins by breaking the entire transcript down into its episodic boundaries. As Strauss explains,

> The *initial* type of coding done during a research project is termed open coding. This is unrestricted coding of the data. This open coding is done . . . by scrutinizing the fieldnote, interview, or other document very closely: line by line, or even word by word. (Strauss 1987, 28).

For the purposes of our undergraduate technical writing courses, asking students to create a written transcript of their videos and then to break them down into line-by-line episodes is a bit too resource intensive. Instead, the episodic boundary I ask students to use is the simple comment. I ask students to go through their videos or audios and count every time a participant starts and stops a new comment. In a professional setting like the Usability Testing Facility I direct, we use a piece of software such as TechSmith's MORAE to complete this task. However, I ask students to do it by just listening to the playback of their protocols, pausing the playback at the end of each comment, counting the comment, and then putting the comment into some open category.

As should be clear from the previous discussion of how to take notes during protocol analysis, students begin this coding process with some of their open categories already predetermined. These include recommendations, negative comments, comments about navigation, comments on confusing sentences, and so on. If a student has a comment that doesn't fit into preexisting open categories, they must create a new one. Oftentimes it's the creation of these new and unpredictable categories that leads students to make new discoveries about the ways users use instruction manuals. For example, I required students to conduct research on video instructions that had been developed in CAMTASIA STUDIO and that documented how to complete a series of tasks needed to build websites with ADOBE DREAMWEAVER. Although neither the students nor I predicted it, we discovered we needed an open category that addressed whether or not the procedures being documented by the videos were short or long. Through meticulous coding of the comments, it was discovered that users of video instructions don't like video as a delivery system when they're dealing with a long, complex series of steps. Users complained that they had to watch the entire video and then memorize the long series of steps they needed to complete in order to build their webpages. Or if they didn't watch the whole video, they had to start and stop the video all the time and lost their train of thought.

In hindsight, the students' observations about poor ease of use that results from documenting long, complex procedures in a video are commonsensical. Anyone who's watched a lengthy video with instructions of this type can probably empathize. Nevertheless, if students hadn't rigorously coded and analyzed the data they collected from their protocols, these observations would have never have been made, and their insights would have been lost.

Often open coding of the data isn't enough to reveal patterns and impose order on the information. Axial coding may also be needed. Strauss defines axial coding in this way:

> Axial coding is an essential aspect of the open coding. It consists of
> intense analysis done around one category at a time, in terms of the
> paradigm items (conditions, consequences, and so forth). This results in
> cumulative knowledge about the relationships between that category and
> other categories and subcategories. (1987, 32)

Put another way, axial coding is simply taking an open-coding category that may be too general and dividing it up into several subcatego-
ries. Take for example the open-coding category "recommendations."
Comments coded into this category are recommendations users made
about how to improve the manual the students are testing. However,
there's a great deal we don't yet know about the quality of the comments
in this category. For example, are the recommendations being made
about one feature of the manual, or do the comments address other
aspects? Do most of the recommendation comments involve problems
with the manual's discussion of, say, installation procedures, while only a
small percentage address navigation and page layout? The axial coding
in this case would take the broad open category "recommendations" and
subdivide it into the three axial categories: (1) installation procedures,
(2) navigation, and (3) page layout. Furthermore, by reporting that the
highest percentage of recommendation comments were addressed to
installation procedures, the author of the manual now can prioritize
what revisions to make. Since installation procedures generated the
most consternation from users, the author should probably dedicate
more resources to working on those problems than on others.

### Write Recommendation Report

As a usability testing professional who conducts sponsored research for
industry clients, my primary responsibility to my clients is to provide
them with *actionable* research results. Industry clients aren't willing to
pay if they aren't provided with substantive recommendations on how
to improve the products that have been user tested on their behalf. This
same standard obtains for the recommendation reports I ask students
to prepare on the manuals they have tested. This is an important point
to make to our technical communication students for two reasons.
First, as was argued at the beginning of this chapter, at least part of
the motivation for integrating usability testing into our classes is to
prepare students for the kinds of experiences they can expect in the
workplace. Second, it's likely many of the students in our technical communication courses come from scientific, technical, or engineering disciplines where they have been trained to believe that technical reports

describing the results of empirical research studies must be "objective" and that they should eschew overtly persuasive agendas. We must take this opportunity to educate these students on the inherently rhetorical nature of reporting empirical findings for clients—clients who are anything but disinterested in their results.

When I say the audience for the students' reports are anything but disinterested, I should explain that I don't assign the authors of the manuals to conduct usability studies on their own manuals. Instead, I put students in pairs and have them test each other's manuals. In this way, the client for the usability test is the student/author of the manual tested. From conversations with colleagues who use usability testing in their classes, I realize this approach is unusual. Many instructors ask the student authors to also be the researchers and test their own manuals. These instructors treat the usability testing as though it were a kind of revision heuristic. However, there are several reasons I believe an alternative approach is warranted.

First, the usability testing community is virtually unanimous about the fact that authors and product developers cannot get the critical distance necessary in order to objectively assess their own products. Nielsen writes,

> If you test your own design . . . , you might be less willing to admit its deficiencies. Designers can be too willing to dismiss user complaints or problems as minor or unrepresentative, when in fact the test indicates the need for a deep redesign. Also, designers can get so caught up in their own theories about how users *ought* to behave that they forget to test for cases in which people behave differently. (Nielsen 2007)

The mindset of developers and authors is necessarily directed at different ends than is the mindset of usability researchers. Developers and authors must be concerned about how to make something work and about functionality. Usability testers must adopt a different mindset and must focus on how something *should* work.

Second, as Nielsen's quote above suggests, student authors are going to have difficulty admitting deficiencies in their manuals. Grades are a powerful motivation for students, and they want the instructor to have the most favorable impression of their work. Students have an extremely difficult time, therefore, providing evidence of their manuals' flaws. They know full well that the instructor will read that evidence and may develop a low opinion of their work. For this same reason, they have an even more difficult time providing video clips that show users who are frustrated and annoyed with their manuals. In short, as the authors of the manuals, they have strong incentives not to do a very good job on their recommendation reports.

Third, not only is it difficult for the authors to be "objective" about the usability study, it's difficult for participants in usability studies to give honest and unencumbered feedback. Once participants learn the person conducting the research study is also the author of the manual being testing, they don't use the manual in a "natural" way. As has repeatedly been shown by the Hawthorne effect, participants in a study modify their behaviors in order to improve their scores or please the researcher/observer. In the case of usability studies, they "pull their punches" when expressing their opinions and withhold critical information they believe may hurt the researcher's feelings. This, of course, biases the outcomes and prevents even an "objective" researcher from making important insights into potential problems with a manual.

Finally, one of the main reasons I don't want the manuals' authors to also be the usability researchers goes back to the point I made earlier about providing students the opportunity to write about empirical research findings for *interested* clients. I find that most undergraduates in my technical writing classes have written lab reports in their chemistry, biology, or electrical circuits courses on empirical "exercises" where both they and their audiences are disinterested in the outcomes. Unless they've had an exciting internship or co-op experience somewhere, writing the recommendation reports for their peers is often the first time they've had an opportunity to write authentically about empirical research they have conducted. I believe it's important to give them that opportunity in a technical communication course.

Teaching students to structure the reports on their usability studies as recommendation reports rather than "objective" technical reports, and teaching them to support their recommendations with actual video clips the manuals' authors can watch, is a win-win. Students learn the importance of providing actionable results in a technical document. Also, through the videos, the authors of the manuals are able to experience in a very personal way how actual users of their writing respond to their writing.

## CONCLUSION

Usability testing is a mainstream skill set for today's technical communicator. It's no longer the fringe activity of a few elite specialists. Instructors in today's technical communication classrooms must focus explicitly on integrating user-centered design processes and usability testing into their assignments. As a result, this chapter focuses explicitly on illustrating how to integrate usability testing into students' document-development processes. Specifically, we've overviewed the five basic phases instructors must go through in the usability testing assignment, and we've shown how to use existing technical writing assignments like instruction manuals and producing documentation sets to integrate usability testing into a course's curriculum. Also, since experience has shown that the most difficult phase of the process for both students and new instructors is learning to code the data collected in ways that lead to *actionable* recommendation reports, I've given specific attention to coding and analyzing the data in this chapter. Of course, no single chapter in an anthology can provide everything you need to know to help students learn to conduct usability tests. Fortunately, as Steve Krug has famously written, usability testing isn't rocket surgery. And so, I hope that as a result of this discussion, along with the resources I listed in the "Getting Started" section above, you'll be encouraged enough to find ways to incorporate usability testing into your technical communication courses.

### DISCUSSION QUESTIONS

1. What kinds of technical writing documents and course assignments lend themselves to usability testing procedures?

2. How can instructors raise students' awareness of usability testing and convince them that the research methodology is worth their time to learn?

3. What are strategies teachers can use to evaluate the processes students use as they are conducting their usability studies?

4. What techniques can teachers use to make sure students are developing a sense of audience awareness from both the usability tests they conduct and the results they receive from other students in the class?

5. What role does coding play in the testing process and how does it relate to what students are learning about writing technical documents?

## APPENDIX 11.A

### *Sample Informed Consent Form*

#### INFORMED CONSENT AGREEMENT

**Purpose**: The students and faculty of [insert your class here] are asking you to participate in a study of the [insert your study name here]. By participating in this study, you will help us make the system easier to learn and use.

**Study Environment**: The study will take place in [insert location], where you will be observed as you use the manual. By signing this form, you agree to abide by the rules of the [location name]. We will be glad to provide a copy of these rules for your review.

**Information Collection**: We will record information about how you use the [product name]. We will interview you before and after your work with the [product name]. We will use the information you give us, along with the information we collect from other participants, to recommend ways to improve the [product name].

**Audio/Video Waiver**: All or some of your work during this study and the interviews will be recorded on [either audio or video]. The researchers shall own the results of the services you perform under this agreement. By signing this form, you give your consent to us to use, reproduce, and publish your voice and verbal statements, *but not your name*, for the purpose of evaluating the [product] and showing the results of our study.

**Comfort**: We have scheduled breaks for you, but you may take a break at any other time you wish. Merely inform the study administrator that you would like to do so.

**Risks**: There is the possibility that an excerpt from an A/V recording of you could be used in a public setting. [Your institution] cannot be held liable for any injury you may receive as a result of your participation in this study.

**Freedom to Withdraw**: You may withdraw at any time.

**Freedom to Ask Questions**: If you have any questions, you may ask either study administrator, now or at any time during the test.

**Compensation**: Your participation in this study is completely voluntary. There is no monetary compensation for participating.

If you agree with these terms, please indicate your acceptance by signing below.

Signature:_____

Date: _____

Printed Name:_____

Your Initials: _____

Witness:_____

Date: _____

Investigator:_____

Date: _____

Participant Number:_____

If you have any additional questions or wish to report problems, please contact [instructor's name] at [instructor's phone number].

## APPENDIX 11.B
### *Sample Video Release Form*

#### VIDEO RELEASE FORM

The signature below indicates my permission for the [class name] to use video footage recorded during the usability session conducted for:

[title of usability study] on _____(date)

in which I served as a participant.

My name will not be reported in association with session results nor will my name be included on the video footage. This video footage may be used for the following purposes:

- Analysis of research and reporting of results
- Conference presentations
- Educational presentations
- Informational presentations

I will be consulted about the use of the video recording for any purpose other than those listed above.

There is no time limit on the validity of this release nor is there any geographic specification of where these materials may be distributed.

This release applies to video footage collected as part of the usability session listed on this document only.

I have been given a blank copy of this release form for my records.

| Name (please print): | Date: / / |
|---|---|
| Signature: | |
| Address: | |
| Phone: | E-mail: |

*References*

Barnum, Carol M. 2011. *Usability Testing Essentials: Ready, Set—Test.* Burlington, MA: Morgan Kaufmann.

Bias, Randolph G., and Deborah J. Mayhew. 1994. *Cost-Justifying Usability.* Boston, MA: Academic.

Bias, Randolph G., and Deborah J. Mayhew. 2005. *Cost-Justifying Usability: An Update for the Internet Age.* Amsterdam: Morgan Kaufmann.

Casey, Steven M. 1998. *Set Phasers on Stun: And Other True Tales of Design, Technology, and Human Error.* Santa Barbara, CA: Aegean.

Casey, Steven M. 2006. *The Atomic Chef: And Other True Tales of Design, Technology, and Human Error.* Santa Barbara, CA: Aegean.

"Certification." Society for Technical Communication. "Certification." Accessed April 13, 2016.

Dumas, Joseph S., and Janice Redish. (1993) 1999. *A Practical Guide to Usability Testing.* Exeter, UK: Intellect Books.

Ede, Lisa, and Andrea Lunsford. 1984. "Audience Addressed/Audience Invoked: The Role of Audience in Composition Theory and Pedagogy." *College Composition and Communication* 35 (2): 155. doi: 10.2307/358093.

Grant-Davie, Keith. 1992. "Coding Data." In *Methods and Methodology in Composition Research*, edited by Gesa Kirsch, 270–86. Carbondale: Southern Illinois Press.

Gurak, Laura, and John Lannon. 2010. *Technical Communication.* New York: Pearson.

Harner, Sandi, and Anne Rich. 2005. "Trends in Undergraduate Curriculum in Scientific and Technical Communication Programs." *Technical Communication* 52: 209–20.

Krug, Steve. 2010. *Rocket Surgery Made Easy: The Do-It-Yourself Guide to Finding and Fixing Usability Problems.* Berkeley, CA: New Riders.

Meloncon, Lisa, and Sally Henschel. 2013. "Current State of U.S. Undergraduate Degree Programs in Technical and Professional Communication." *Technical Communication* 60 (1): 45–64

Miller, Carolyn R. 1989. "What's Practical about Technical Writing?" In *Technical Writing: Theory and Practice*, edited by Bertie E. Fearing and W. Keats Sparrow, 14–24. New York: MLA.

Nielsen, Jakob. 1995. "Severity Ratings for Usability Problems." Nielsen Norman Group.

Nielsen, Jakob. 2000. "Why You Only Need to Test with 5 Users." Nielsen Norman Group.

Nielsen, Jakob. 2007. "Should Designers and Developers Do Usability?" Nielsen Norman Group.

Rose, Mike. 2009. *Writer's Block: The Cognitive Dimension.* Studies in Writing and Rhetoric. Carbondale: Southern Illinois University Press.

Rubin, Jeff, and Dana Chisnell. 2008. *Handbook of Usability Testing.* 2nd ed. Indianapolis, IN: Wiley.

Sauro, Jeff. 2013. "Rating the Severity of Usability Problems." Measuring U.

Society for Technical Communication. 2013. "Certification Maintenance Policies for Certified Professional Technical Communicator (CPTC)."

Strauss, Anselm L. 1987. *Qualitative Analysis for Social Scientists.* Cambridge: Cambridge University Press.

Tullis, Tom, and Bill Albert. 2013. *Measuring the User Experience.* 2nd ed. Waltham, MA: Morgan Kaufmann.

Usability.net. n.d. "International Standards for HCI and Usability." Accessed March 26, 2018. http://www.usabilitynet.org/tools/r_international.htm.

## 12

# WHAT DO INSTRUCTORS NEED TO KNOW ABOUT TEACHING TECHNICAL PRESENTATIONS?

Traci Nathans-Kelly and Christine G. Nicometo

### WITNESS THE SLOW DEATH OF A GOOD INSTRUCTOR

You are an instructor of technical communication. You teach three sections of this course. It is the end of the semester, and students are required to give a talk of about fifteen minutes on a technical issue. That is 75 talks you will watch and critique.

You know you have a bevy of brilliant students talking about complex, exciting work. The days march on, and you watch seventy-five presentations in which students have used slides. Each slide has a fragment as a header and three to nine bullets of text. If you are lucky, sometimes a graph shows up on the slides to break up the monotony.

The students are nervous: they read off their slides, they fidget, they hide behind a podium, and they have poor eye contact. Also, in each presentation, the point—the very claim and core of their work—is lost in the sea of bullets.

After about the first ten talks, your eyes glaze over. You see the other students in the room tune out. They are looking at their computers or their phones. Maybe some are dozing. You need to evaluate the speakers' work, but you cannot focus. You wonder why (why?!?) the talks are so bad and why you sit through this every single term. You die a little on the inside. As you drag yourself from the classroom each day, you know there must be a better way.

### ASSESS THE POOR TECHNICAL PRESENTATIONS OF OLD

If you are going to require presentations of engineering, science, or technical students, you are in for an interesting ride. Although the practices have been shifting a bit in the past few years, these types of presentations are steeped in longstanding, ill-formed traditions that must necessarily be

DOI: 10.7330/9781607326809.c012

challenged by you—the communication instructor. You must challenge not only your students but also yourself and other engineering or science instructors. Your efforts, however, will bring amazingly positive change.

The genre of technical presentations holds many subgenres: the design review, the recommendation, the proposal, the pitch, the status report, the budget review, the training session, the pure information transmission, and the design review (to name a few). The basic techniques we discuss herein apply to all those subgenres, but the speaker must do the necessary work of determining audience, purpose, timing, and persuasive strategy. That task must be done, anew, each and every time the talk is given.

But let's back up a bit and try to understand the problem from a particular point of view. Many students in STEM fields have been told from a very young age, maybe early middle school, that they are talented in the technical fields and that they will not have to worry about the other skills, like communication, very much—if at all. They are told this by teachers, advisors, and parents alike. They are told, for years, a mythology that their technical work will speak for itself and that they will not have to fuss about, communicating it well. Such advice is a terrible disservice to these students, who will be communicating every day in complex circumstances, relaying difficult concepts to varied audiences who all have diverse motivations for accepting or denying what these students will be telling them.

All these factors can have an impact on an unsure young college speaker in a technical communication course. On the whole, technical presentations of all kinds tend to be fraught with many poor habits. As part of that sad set of habits, slides are often to blame. Whether it is fair to assign the blame to the slides or the presenter who created the slides, the fact is that quite a few problems can be solved with a retooling of slides' use.

To be honest, slides are a boon for technical presentations because speakers can more easily show technical drawings, pictures, and graphs that often capture and relay their work well. However, the downfall of slide use in the technical and scientific fields is that speakers have also come to use slides as teleprompters. People put too much text or data on the slides, and then an almost-unavoidable action takes place: speakers don't look at their audiences (instead, they look at their slides). Speakers put too much data on each slide, they eschew the core task of doing a good audience assessment, and they tend to make presentations about what they want to talk about rather than what the audience wants or needs to hear.

Another factor that has encouraged poor slide use is organizationally bound; some professors and managers have begun to accept slide decks/files rather than traditional reports as documentation files. If a report is written, sometimes the slides reflect the report, but sometimes the slides *are* the deliverable. When this is the case, the tendency to use too much text on slides is yet again reinforced. We must help students and practitioners alike understand that slides must support the live talk and also function as an archival document. This can easily be done. Read on.

For almost a decade, we have been driven to change up these habits and poor practices as they now stand in the technical fields. To that end, we wrote a book called *Slide Rules: Design, Build, and Archive Presentations in the Engineering and Technical Fields* (Nathans-Kelly and Nicometo 2014) that specifically addresses many of the issues and concerns technical experts have about their talks. We are not alone in this work, as other colleagues such as Michael Alley and Jean-luc Doumont are also making similar calls for retooling technical-presentation design (Alley 2013; Doumont 2009). However, despite the growing voices in this field, poor practices still abound and must be challenged if we hope to nurture communicators and not simply slide readers in our classes.

## KNOW THAT STUDENTS MIMIC THE POOR
## TALKS THEY HAVE WITNESSED

If students give poor talks, it's not necessarily their own fault. They have witnessed many bad presentation and speaker habits over their years of schooling. And, because they are good students, they have internalized these performances as examples of expected behaviors.

All too often, less-than-great talks are because of negligent slide practices. Students see instructors in high school who may use the publisher-provided slides that are nothing more than index cards of notes put into slide form and projected on the wall. Students are subjected to instructors who have transcribed their notes, normally housed on paper and sitting on the podium for personal reference, into slides; as such, those scratch notes are now also projected onto large screens. Instructors treat the prescribed bullet-ridden templates as gospel, diligently filling in bullets and thinking this will make for a great talk.

These inclinations are amplified in the technical areas, where experts and/or instructors and professors believe every single technical point must be written out on the big screen in order for others to understand (see fig. 12.1 for a typical bad slide).

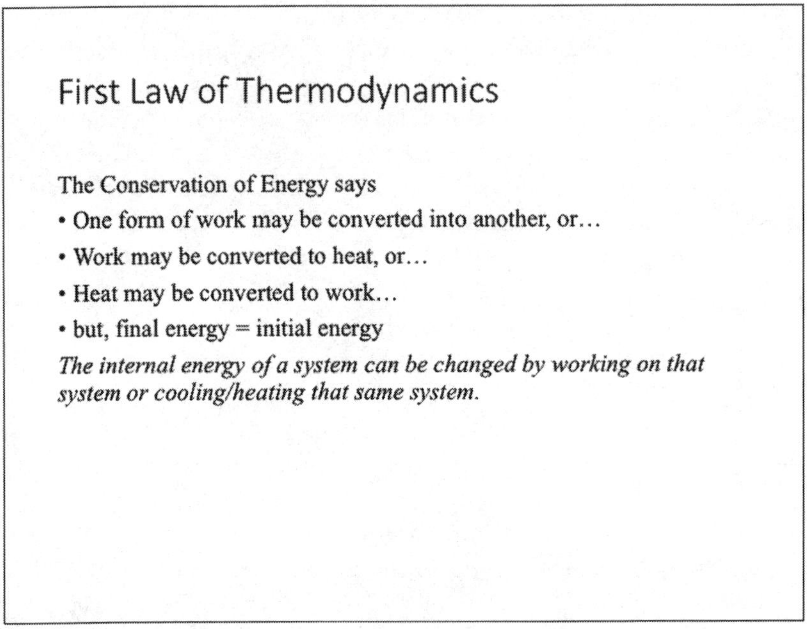

## First Law of Thermodynamics

The Conservation of Energy says
• One form of work may be converted into another, or...
• Work may be converted to heat, or...
• Heat may be converted to work...
• but, final energy = initial energy
*The internal energy of a system can be changed by working on that system or cooling/heating that same system.*

*Figure 12.1. Typical poor slide. The impulse to combine into one file both the technical information and full documentation looms large for many speakers. However, slides like this actually demote the power of the expert who is speaking at the moment, which we want to avoid.*

Aside from seeing poor practices in the past, there's an understandable method to this madness of too many bullets, and it is this: speakers want to provide documentation and information, and they want to make just one file for efficiency's sake. Indeed, to combine these needs into one file is reasonable. However, the execution of that need has been pretty dismal and leads to terrible live presentations. There is a way out of this mess, however.

## TEACH TECHNIQUES THAT EMPOWER
## AND INFORM

We maintain that despite years of seeing poor practices, technical presentations can be remedied in pretty quick order. Students are smart, and if they are asked to reconceive how slides are used, they are able to do so. When you can explain that slides, like all other technical or scientific work, are enhanced by a systems way of thinking, improvements naturally come. Engineering, science, and technical students are well versed in processes and design work; they can apply logical methods in an effective and thoughtful manner. Your job, as the communication instructor, is to demonstrate new processes and how to apply them, strategically, to their various speaking tasks.

To begin, address the smaller items that often plague STEM students and make them bemoan giving talks: being nervous, not knowing what to do with their hands, not knowing where to stand—these are the bits and pieces that seem mysterious to them.

In addition, by framing the new slide techniques we discuss here as proven, effective, dual-purpose files that can fill several task orders at once (live-presentation support, keeping the audience focused, and fully realized technical notes), students can be convinced that the ways they have seen talks done in the past are not optimal. They come to see that those weak forms can be tweaked for powerful outcomes.

### ADDRESS STUDENTS' FEARS OF SPEAKING

Inexperienced speakers are often nervous about every aspect of public speaking, and we must remember to address these issues during the course of the semester in order to bring some sense of ease to their performance.

#### "I'm Nervous."

Because of the "tracking" of students into STEM courses, it is likely many students do not have substantial experience giving talks. They are nervous, sometimes to the point of making themselves ill. Working with this discomfort can be a bit difficult because nerves lessen with practice, and practice makes students nervous. But we have found several ways to help them think about this.

The first is to have a discussion in class about the performance of the talk itself. No speech in front of a group is ever natural; it is a prepared, staged, and scripted event and it must be treated as such. They must put on the correct costume (likely a "business-casual" outfit on the day of their talk). They must have thought about their words. They must put on the "character" of practitioner/expert in their subject. Amazingly, talking to students about the staging and the character of a subject-matter expert allows them to gain some distance from the audience observers. They feel more protected and more empowered at the same time.

You can also have a candid conversation with students about nerves; being nervous is good. If you aren't a bit nervous, maybe you are too confident and have lost your edge. Nerves keep you alert. In a way, they can be quite helpful.

Another way to empower students is to discuss the power of the moment. When people are gathered in a room to hear someone speak, it is an important moment. An audience has collected itself for a purpose—to hear the speaker. That speaker should be glad for the moment, for there are very few moments in a lifetime when such opportunities arise. Encourage students to harness the energy and shine with their expertise rather than feeling pressured. It's easier said than done, but it's a perspective they may not have heard before.

To take this perspective even further, you can talk to them about the return on investment for them as speakers. For their classmates or for their imagined on-site work colleagues, ask them to roughly calculate an hourly wage. Then ask them to add up that money for one hour for everyone in the room. If you are asking them to imagine future work colleagues, ask them to think about whether or not colleagues have been flown in to hear their talk and to calculate in those costs, too.

At first, this may take students aback, making some nervous again. Your job is to recast that emotional reaction. Talk them through this reaction, helping them understand that people are in a room to hear them talk about their topic. This topic has never been discussed in just this way, for just this purpose, ever before. Empower the students to see themselves as subject-matter experts, even for just those fifteen minutes. Whatever investment the audience or organization has made to gather in that space to hear the student speak must be worth it and show a strong ROI (return on investment). They must create a talk that is worth that dollar amount since they are the subject-matter experts (SME). They must choose whether to waste the investment by focusing on themselves and their inner dialogue of nerves and fears or to use the investment well by living up to the trust the audience has given them during the time they present. We have found this exercise brings comfort, focus, and a sense of purpose to late undergrads, seniors, graduate students, and practicing experts alike.

### "I Don't Know What to Do with My Hands."

Not knowing what to do with their hands is often a concern for inexperienced speakers. In addition, in the technical fields, the lore is that appearing casual means you look more in control and more approachable. Students interpret this as permission to put their hands in their pockets, which is a terrible habit. Hands in pockets are hands too near inappropriate places in many cultures. Also, pockets often have coins or keys in them, and many speakers begin to play with those items as they talk. Hands in pockets also means the speaker is likely slouching a bit, constricting lung capacity and hindering breathing.

The solution is an easy one: Reagan arms. Yes—the past US president, Ronald Reagan, was a trained actor and politician, but for several years, he was also a spokesperson for GE. He learned quite a bit as he created bridges between technical experts and the general public.

When you can find footage of President Reagan away from his podium, you will see a great technique for controlling hand and arm movement. Model this for students: standing tall and comfortably straight, knees not locked, arms bent at the elbows. With the elbows anchored loosely at the sides, the hands come together in front of the belly. This beginning stance, with elbows at the sides and hands in front, is a good starting range for arm movement. (When explaining this to students, ask everyone to stand up and take this stance; it makes for a fun moment because everyone begins to do robot or t-rex moves and laugh a bit. They will never forget it!)

With arms in this starting position, a speaker can gesture smoothly within a controlled range. If the speaker needs to point to something on a slide or a blackboard, the arm can go out and come back to the starting position. Reagan arms center the hands, simply. For the fidgeters, we find that modeling options also helps: show students how they can subtly play with a ring on a finger, or pass a paperclip quietly between the index fingers and thumbs, or hold onto a pen (not a clicky one) like handlebars. All these "centering" tricks show students tools for controlling their hands. Do not allow laser pointers because those introduce a mess of other problems.

They may balk at Reagan arms at first, saying it feels unnatural. They will laugh as you make them stand up and practice as a group in class. You can point out that the best speakers use this hand-controlling technique (ask them to look at almost any speaker in the TED-talk roster; see fig. 12.2). Actually, this position becomes pretty natural in short time for the students. Let them know this is also a "safe" place for hands to be in all cultures, which is an important consideration for public speaking.

Another advantage to this starting position is that it naturally makes the speaker stand straighter with the head higher. The chest is upright and open, creating the maximum lung capacity. This, in turn, allows for breath regulation, variance in cadence, and control of volume.

*Figure 12.2. Controlled movement with Reagan arms. Using the Reagan-arms tech-nique can provide a natural, comfortable range of motion for speakers. Look to almost any TED talk to see the practice in play. Here, we see technical experts (left to right) Patience Mthuzi, Fred Jansen, and Kenneth Shinosuka giving powerful TED talks, con-trolling their arms and hands beautifully (Jansen 2015; Mthuzi 2015; Shinosuka 2014).*

*"Where Should I Stand? Why Can't I Use the Podium or a Laser Pointer?"*

We encourage speakers to get out from behind the podium because it is often used as a barrier between the speaker and the audience; as such, it prevents a speaker from connecting to that audience. Of course, there are rare moments when speakers at larger events must accept that they must be somewhat tethered to a podium because of microphone issues, but those instances are few and far between.

Instead, speakers should own the space that has been given to them in front of the room. They should be encouraged to move around, in a controlled and professional manner, within that space while speaking and facing the audience. There's power in using the whole area to your own advantage, and students should learn to harness that power for themselves. Habits to avoid include pacing and swaying, and we have found that video recording student talks and requiring them to watch themselves goes far in abating these habits.

If speakers use slides, it is also to their advantage to get out from behind the podium and interact with the slide. There is no harm in moving into the slide to point out an important bit and then moving out of the projec-tion. (For perhaps the most famous example of this, ask students to look at Hans Rosling's TED talk at http://www.ted.com/talks/hans_rosling_shows_the_best_stats_you_ve_ever_seen (Rosling 2006). The magic begins around the 3:15 minute mark.) Using the body (by pointing or gesturing) is always better than a laser pointer, which is hard for many audience mem-bers' eyes to track. Also, a laser reveals shaking hands, and it cannot be "caught" well when recorded. And because so much of the world's work is done now virtually, laser pointers are useless in those moments.

### "Where Do I Look?"

Besides not knowing what to do with hands or where to stand, students worry about eye contact. We give them two techniques to try out.

The first is to imagine a box around a person's head—shoulder to shoulder and up just past the top of the head and over. This area is where you look. It is professional and safe and easy. Within this box, there are also the points of the eyes and just above bridge of the nose. These are stopping points. With some subtlety, the speaker can look at any of these points and still be considered to be making eye contact. Having these stopping points encourages speakers not to stare at anyone straight in the eyes, which is uncomfortable for both.

The second piece of advice we like to share is for the whole room. Visually connecting with each individual in the room is important, as it is a strong method for audience engagement. Thus, encourage students to use the "sprinkler" idea; sweep the room with eye contact. You don't want any part of the room to be left out—to wither and die. Again, much like Reagan arms, this simple visual is something the students remember and carry forward. (Note: a discussion of the "norm" of eye contact in other cultures is a good one to have, as many cultures have extreme interpretations of whether eye contact is respectful or not.)

It is good to introduce these easy tricks and processes to students and then reinforce them all during the class. Addressing these small anxieties makes them a bit more comfortable when speaking because they feel they now have some tools to help them improve. However, we have found that the most momentous change comes when we start talking about their slides.

### TEACH AGILE BEST PRACTICES FOR SLIDE USE

By far, the most revolutionary elements of presenting we teach have to do with slide creation. When people deploy the techniques we promote, their talks become much stronger overall. Indeed, slides have become a crutch, and too many speakers equate a slide deck with a speech. It is a sad state of affairs.

So the issue of slides must be dealt with, and you must have a strong stance and plenty of answers for the questions that will undoubtedly arrive, from students, colleagues, and engineering practitioners alike (Nathans-Kelly and Nicometo 2011).

Ill-conceived slides encourage many bad habits: using a monotonous voice, turning the back to the audience, not preparing/knowing the content, using fragmented thoughts, reading off the slides, and conflating the slides for the live talk with the in-slide documentation of the topic. If you introduce and encourage (require?) the techniques we provide, the slides will become a strong background for the speaker. Our entire purpose is to once again uphold the *speaker*, not the slides, as the most important part of the talk.

There are three main elements to this new slide approach:

1. Headers should be short, complete sentences.

2. The main slide acreage should hold a visual.

3. Heavy text, bullets, notes, citations, and other word-heavy elements belong in the notes pane.

These changes in slide construction can make an immense difference in how students understand their own content, how they formulate the "story" of their technical work, and how they interact with their audiences. Michael Alley championed the first two elements for years with his own research and teaching (Alley 2013, 2011; Alley et al. 2007; Alley and Neeley 2005; Alley and Robertshaw 2004; Alley, Schreiber, and Muffo 2005; Garner et al. 2011). After working with practicing engineers, scientists, and technical experts, we added number three to meet the organizational needs of technical work (Nathans-Kelly and Nicometo 2014) along with other needs noted by professional engineers. If you are teaching students who are more into marketing or business, other books might suit your purpose better (Atkinson 2011; Duarte 2008; Reynolds 2011). Of course, adaptations to suit particular needs are always part of the game.

## ENCOURAGE AND SUPPORT THE TRANSFORMATION

Here is an encouraging word: we find that when speakers use the slide practices shown in the next few pages, they feel a tremendous relief, for they are free to speak with expertise on their topic in their own voices. The techniques provide the mental room and the necessary visual backup for the speaker to absolutely shine. You will see speakers regain their confidence, engage authentically with the audience, and speak with ownership on the topic. The transformation can be stunning.

They key to the new slide approach is this: empowerment. Students should feel empowered to take the stage, talk through their complex topics, and answer questions confidently without feeling as if they are playing second fiddle to their own slides (Nathans-Kelly and Nicometo 2013). Let's get started.

*Headers Should Be Short, Complete Sentences.*
Start changing the slide paradigm right at the top. As we saw in figure 12.1, a usual feature is a fragment as a header. In truth, a fragment is just a keyword for the speaker and isn't of much use for the audience. A fragment means the audience now must scramble to try to make a complete thought of the fragment; in doing so, mental energy is moved away from the speaker and into solving a puzzle. This is not a desirable outcome, especially during a technical talk in which the speaker may be saying something out loud that is essential. For the audience, a fragmented header is a quick invitation to distraction.

Instead, a short (no more than two lines) complete thought is a better move. A sentence header (a statement, not a question) denotes authority, control, and focus. Interestingly, as speakers work at writing these short sentences for their slides, they gain control of their topic and a focus for their technical story. Organizationally, their talks become stronger because of this simple change. The purpose of the talk is honed by focusing on the main ideas in this manner.

A sentence header also means only *one main idea can exist on a slide.* Does this promote more slides in a deck? Maybe so. However, it does not mean the talk will be any longer; it means that as a visual, each slide is focused and useful for the audience. *Do not give students slide counts as a constraint. Give them time as a constraint and clarity as a goal.*

With our thermodynamics example again, consider this retoooling in figure 12.3 (we are just looking at the header for now). Instead of just proclaiming there is a law, as in figure 12.1, we have now given a bit more information. There is a complete thought that can be unpacked. The audience doesn't have to play "fill in the blanks" with a fragmented header; the thought is fully realized, allowing the audience to quickly read it and then refocus on the speaker.

Furthermore, the sentence header works extremely well as a prompt for the speaker. Again, fragmented headers could flummox a nervous speaker who cannot remember what the "real" or more elaborate main point was. The speaker should not read that header word for word to the audience. Rather, it should be a prompt, helping the discussion to move forward.

Let's look at one more example to see how a change in header can promote true information transfer (fig. 12.4).

In the past, we have all seen slides that have one word headers: *Results, Analysis, Recommendations, Materials,* and so forth. While brevity can be a virtue, too much brevity is actually a hindrance. To ensure a memorable takeaway for the audience, be forthright and make a statement. The

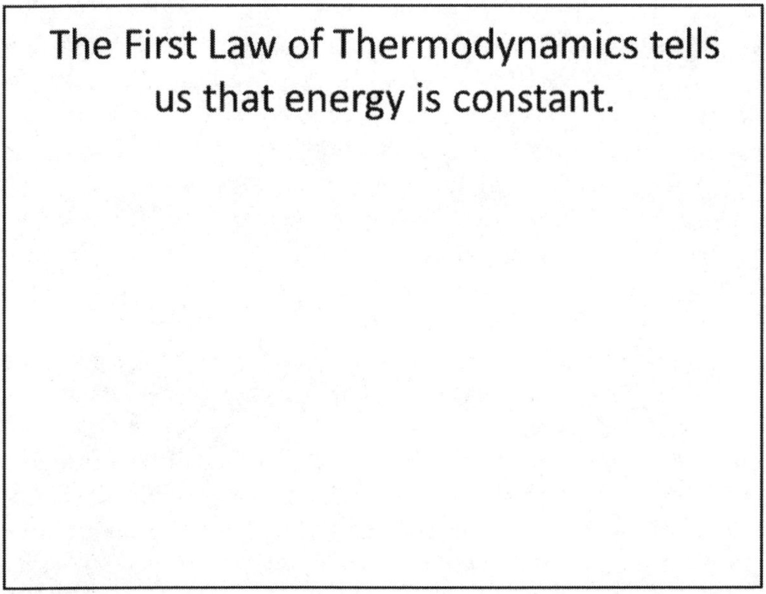

> # The First Law of Thermodynamics tells us that energy is constant.

Figure 12.3. Encourage the use of sentence headers. Short, efficient one-topic headers work much better than fragments for framing complex technical matters.

audience will be thankful and have a much better chance at retaining the major points.

*The Main Slide Acreage Should Hold a Visual.*

After the header has become engaging, so should the rest of the slide. The change here is to refocus the purpose of the slide's' main acreage away from being a teleprompter for the speaker into something more meaningful for the audience. For some, that is a major paradigm shift— moving the slide away from being a large, glowing index card into a visual that is powerful and meaningful for the intended audience.

This change necessarily means bullets must be trimmed and even eliminated. Before panic sets in, please understand that those bullets will still be housed with the slide deck, but the bulleted information will reside in the notes pane (see the next section for more details). For a live talk, remember, the focus should be on the expert at the front of the room. The talk is not the slides. The slides are a prop and should never overstep their purpose.

Another key factor is that when we simplify the slides, we lessen the chance of cognitive overload for the audience. No matter how smart

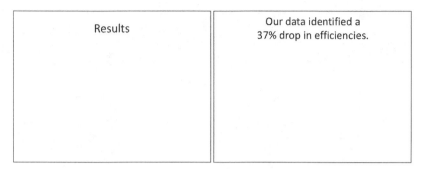

*Figure 12.4. The old and the new style of header. On the left, the one-word header leaves an audience without much to work with. On the right, we have a short statement that is forthright and complete; it works well to frame the major point the speaker wants to make.*

audience members may be, it is a difficult task to listen to a speaker, understand the speaker, read a slide while listening, understand the slide, and knit all that information together in quick order. It simply is too much to ask of anyone and expect the information to be retained. When the slide for the live talk is cleaned up and well constructed, you do an incredible favor for anyone at the talk.

Thus, to support the speaker, the combination of the sentence header and the on-topic visual delivers a powerful one-two punch for the audience. Forget cute clip art and meaningless cartoons. Technical presentations are not the place for such silly work. Make a practice of using visuals that will support the technical points and provide a mental anchor for the concepts. See figures 12.5 and 12.6 for examples. Complex content can be just as easily presented as more simple concepts (see fig. 12.7).

In addition, encourage good critical thinking about these changes. What information does the audience expect? What does a person need who is looking only at the archived slides and not seeing the live talk? How does the slide deck function as a living document inside an organization? Is it just talking points? Is it a reference document? Knowing the answers to these kinds of questions about larger organizational issues allows any speaker who is using slides to tweak these techniques to find a unique approach specific to the task and technical topic at hand.

*Use the Notes to Solve the Too-Much-Text Issue.*
From the very first version of slide software, the creators had a great idea: the notes feature. In PowerPoint® and Keynote® alike, the speaker

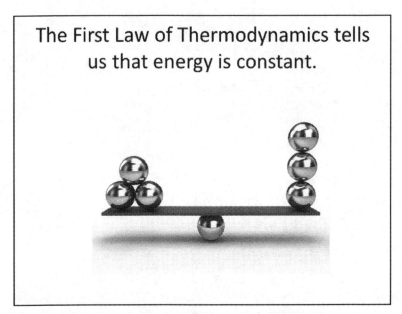

*Figure 12.5. Targeted visuals reinforce the speaker's point. Using the main area of the slide, visuals can bring another layer of meaning to the speaker's intended message.*

*Figure 12.6. Build a case. Using sentence headers and visuals, you can build your point, bit by bit, without overwhelming the audience with a cluttered slide that holds too much information at once. Give listeners time and space to take in new ideas, building on them at a reasonable pace.*

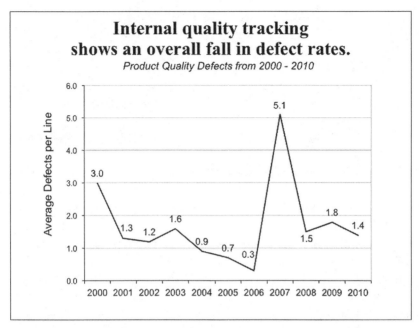

*Figure 12.7. Complexity can be clear. Using sentence headers, clear visuals (even when complex) for targeted audience needs can allow technical information to shine on slides. See figure 12.8 to understand how to use notes for archival advantage.*

can transfer much of the text off the slide itself and into the notes. There are several key advantages to this practice:

1. The slides become less cluttered, lessening cognitive overload.

2. The slides support the speaker better in the live-talk moment.

3. The slide deck is richer for using the notes because it can contain more than the slide.

4. A PDF of the slide deck as documentation is a richer document than slides with no notes at all, contributing to its function as a kind of report or archival piece.

Let's briefly explore these advantages of using the notes feature.

*The slides become less cluttered* when the bulk of the text we often see in bullets gets transferred to the notes pane. By cleaning up the slide for the speaker's actual talk, we do a favor for the live audience. The notes pane can house bullets, long sentences, citations/references, and other speaker notes. Also, if a talk asks for interaction and feedback from the audience (like a design review), the speaker can record information right in the notes pane during the session itself. This move is incredibly powerful because the input from others in the room now becomes part of the slide deck itself—a living record of information exchanged during the meeting between stakeholders.

*The slides support the speaker* better because they are not laden with bullets or other heavy text. With a sentence header and a great visual alone, speakers are free to showcase their expertise on the topic without trying to stay lockstep with the slide's wordy content. As an instructor, you will marvel at how well this works for students when they are allowed to shine instead of reading a teleprompter-style slide.

*Slide decks also become rich repositories* when the notes are used because they can house more than any slide alone ever could. From technical content to citations to speaking points to audience input, the notes can hold almost unlimited amounts of additional information that may be spoken aloud by the speaker or needed later when the slides are functioning as more of an archival piece or documentation of the technical work.

To that end, a *PDF of the slides, when requested, is our preferred way to send slides* to anyone who has missed the talk or for record keeping for a project. For students and practitioners alike, a PDF is easy to open on almost any device and is relatively stable in its format. The trick to this is to create a PDF in the right page format. When making the PDF, choose to show the notes as part of your print/save options. In this way, when the recipient opens the document, both the slide content and the notes content will be shown equally (see fig. 12.8).

This is a powerful move on your part, as the sender, for it doesn't allow anyone to "miss" the notes if they are viewing in the slideware program. A PDF, properly created, solves this problem.

## CONCLUDING THOUGHTS

Moving toward a more targeted, useful, and rich use of slides to support technical talks is a journey, indeed. First, model the practices yourself. Ask that students take on the challenge; they will do remarkably well and thrill you with how well they come to speak about their technical expertise. These methods are tried and true, and they adjust well to meet organizational needs, specific course requirements, or other imposed constraints.

Remember, too, to let any other faculty you are working with know about the updated techniques you are teaching students so they are not caught unawares. In the best circumstances, other faculty may not notice anything more than improved talks. In other cases, the change to the codified slide practices may concern them, and they will need to know about the use of notes as the bucket for the old bullets.

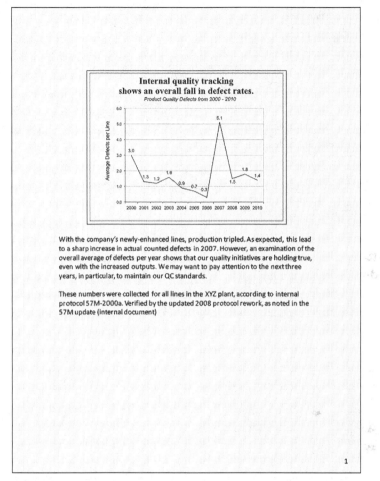

*Figure 12.8. Archival work with slides can be rich. Using the notes, and then sending slides as PDFs with full notes showing, is a powerful way to transmit both the slide's contents and the full notes.*

Carry on teaching students the staging tricks, the breathing techniques, and Reagan arms. And insist on those slide changes, for they will revolutionize how each and every speaker thinks about stage versus archival needs, forever. They will thank you.

## DISCUSSION QUESTIONS

1. In your school, college, or university we encourage you to seek out, investigate, and discuss established "tradition" for using slides at your school, college, or university for

a. instructional materials

b. required assignment artifacts from students.

After reviewing these established institutional materials, where can you identify opportunities for meaningful change in practice? How can you prioritize those opportunities for change?

2. In your assessment, in what way are those traditions supporting good presentation practices? How could they do better?

3. Changing communication patterns and habits (slides or otherwise) can lead to challenging some fundamental ways teachers perceive and frame their work. In altering the "old style" of slides, how would you be challenging other instructors to reframe their pedagogical stances?

4. When a teacher insists that extra text, notes, citations, and other supportive materials be housed in the notes pane, how do slides transform information transfer? How could this simple move be empowering for teachers and students (and eventually working professionals)?

5. Students will encounter other presenters that insist the bulleted style is correct and more desirable. What can you do to help students explain their innovative ways to people who are doubtful?

## References

Alley, Michael. 2011. "Rethinking the Design of Presentation Slides." Paper presented at the IEEE International Professional Communication Conference, 2011, Vancouver, BC.

Alley, Michael. 2013. *The Craft of Scientific Presentations: Critical Steps to Succeed and Critical Errors to Avoid.* 2nd ed. New York: Springer.

Alley, Michael, and Katherine A. Neeley. 2005. "Rethinking the Design of Presentation Slides: A Case for Sentence Headlines and Visual Evidence." *Technical Communication* 52 (4): 417–26.

Alley, Michael, and Harry Robertshaw. 2004. "Rethinking the Design of Presentation Slides: The Importance of Writing Sentence Headlines." Paper 61827, 2004 International Mechanical Engineering Conference and Exposition. Anaheim, CA: Association for the Society of Mechanical Engineering.

Alley, Michael, Madeleine Schreiber, Elizabeth Diesel, Katrina Ramsdell, and Maura Borrego. 2007. "Increased Learning and Attendance in Resources for Geology through the Combination of Sentence-Headline Slides and Active Learning Measures." *Journal of Geoscience Education* 55 (1): 85–91.

Alley, Michael, Madeleine Schreiber, and John Muffo. 2005. "Pilot Testing of a New Design of Presentation Slides to Teach Science and Engineering." Paper 1213, 2005 Frontiers in Education Conference. Indianapolis, IN: ASEE/IEEE.

Alley, Michael, Madeleine Schreiber, Katrina Ramsdall, and John Muffo. 2006. "How the Design of Headlines in Presentation Slides Affects Audience Retention." *Technical Communication* 53 (2), 225–34.

Atkinson, Cliff. 2011. *Beyond Bullet Points: Using Microsoft PowerPoint to Create Presentations That Inform, Motivate, and Inspire.* 3rd ed. Redmond, WA: Microsoft.

Doumont, Jean-luc. 2009. *Trees, Maps, and Theorems: Effective Communication for Rational Minds.* Brussels: Principaiae.

Duarte, Nancy. 2008. *Slideology: The Art and Science of Great Presentations.* Sebastopol, CA: O'Reilly.

Garner, Joanna K., Michael Alley, Kerri Lynn Wolfe, Sarah E. Zappe, and Lauren Elizabeth Sawarynski. 2011. "Assertion-Evidence Slides Appear to Lead to Better Comprehension and Recall of More Complex Concepts." Track: T245 Rethinking PowerPoint and Other Acts of Communication, 2011 American Society for Engineering Education Conference, Vancouver, BC.

Jansen, Fred. 2015. "How to Land on a Comet." Filmed March 2015. TED video, 17:48.

Mthuzi, Patience. 2015. "Could We Cure HIV with Lasers?" Filmed March 2015. TED video, 4:26.

Nathans-Kelly, Traci, and Christine G. Nicometo. 2013. "Stop Slipping and Sliding:Methods to Reclaim Expert Engineering Space byUsing Slides to Best Advantage." Paper presented at the meeting of the SEFI/European Society for Engineering Education, Leuven, Belgium, September 2013.

Nathans-Kelly, Traci, and Christine G. Nicometo. 2014. *Slide Rules: Design, Build, and Archive Presentations in the Engineering and Technical Fields.* Hoboken, NJ: Wiley-IEEE.

Nicometo, Christine, and Traci Nathans-Kelly. 2011. "Informed Influence: Preparing Graduate Engineering Students to Present with Power and not Just PowerPoint®." Paper presented at the American Society for Engineering Education Conference, Vancouver.

Reynolds, Garr. 2011. *Presentation Zen: Simple Ideas on Presentation Design and Delivery.* 2nd ed. San Francisco, CA: New Riders.

Rosling, Hans. 2006. "The Best Stats You've Ever Seen." Filmed February 2006. TED video, 19:46.

Shinosuka, Kenneth. 2014. "My Simple Invention, Designed to Keep My Grandfather Safe." Filmed November 2014. TEDYouth video, 5:47.

## 13

## TEACHING INTERNATIONAL AND INTERCULTURAL TECHNICAL COMMUNICATION
### *A Comparative Online Credibility Approach*

Kirk St.Amant

It's presentation day, and the students in your class will soon share web designs with the community partners you assigned them at the start of the semester. At first, all seems to be going okay, but then one of your students encounters problems. The community partner—the director of a volunteer organization working with immigrants from Central Asia—does not like the design developed for this project. The student, who tested the design with friends and family, is confused, and you're a bit flummoxed too. What went wrong?

To answer this question, you later ask the community partner why the site was problematic. The reasons quickly become clear:

- The information on the site is in English, a language few of the recent immigrants speak or read.
- The images on the site reflect Anglo-American expectations of place (e.g., rooms with objects common in most US homes, but not in those of the intended audience).
- The site was designed for use on a desktop or laptop computer—a technology little used by the intended audience (who instead use smartphones to go online).

The situation makes you keenly aware students need to be introduced to aspects of culture and communication in relation to a range of situations. But the question is how.

The need for training in culture and commutation is certainly there. In today's global economy, new graduates can often find themselves creating artifacts—or informational or instructional materials—for individuals from other cultures. The challenge for instructors is how to provide meaningful educational experiences that can help students understand international communication contexts.

DOI: 10.7330/9781607326809.c013

Addressing this situation is not easy. Technical communication instructors can, however, benefit from heuristics—or models—that help students grasp broader factors affecting culture and communication. Ideally, such heuristics

- guide student research on topics relating to culture and communication;
- provide a foundation for applying ideas in meaningful ways;
- assist students in developing skills for creating materials for other cultures.

This chapter provides technical communication instructors with an example of such a heuristic.

The approach I describe here is founded on the idea that credibility is central to effective international technical communication. I first examine how expectations of credibility and communication arise in a culture. Next, I provide an overview of a heuristic-based approach to teaching students these ideas of culture and credibility in different classes. I then conclude with example assignments instructors can use to introduce students to these ideas. Such an approach can provide students with a foundation for researching, understanding, and addressing cultural credibility expectations.

### LITERATURE REVIEW

Effective international communication encompasses two factors: credibility and usability. Credibility involves whether individuals perceive an item as a legitimate source of information—one worthy of time, attention, and consideration. Usability entails whether individuals wish to use—or are able to use—ideas or information as intended. When introducing students to these ideas, the key is to emphasize that credibility and usability both affect how humans perceive and respond to items they encounter in the world around them.

### TEACHING CREDIBILITY

Perhaps the first step in teaching students about international communication is helping them understand how culture and credibility are (inter)connected. To do so, instructors must explain how—in general—credibility drives use. That is, if something seems credible, you are more likely to pay attention to it and to use it (St.Amant 2006, 3). (To help students conceptualize this point, instructors could have them discuss the aspects they look for to indicate the information provided via certain media—a website, a Facebook post, or a Twitter tweet—is credible.) Once students understand this premise, they must understand another central factor: credibility is not inherent to an item. Rather, our native culture teaches us what attributes or aspects something should have for our culture to consider it credible (Aitchison 1994, 449; Ross and Makin 1999, 205, 209).

Next, the instructor should explain how cultural factors can influence credibility expectations and result in important differences in communication patterns. Instructors can do so by noting the following: from birth on, we are all exposed to different elements in the world around us. Unfortunately, we have no inherent mechanism for understanding what to pay attention to (or not) in that world. Likewise, we do not know how to value, perceive, and interact with our surroundings. Our native culture, in turn, provides the framework for achieving these goals (Holliday 2013, 1–3). The culture into which we are born tells us what in the world around us matters (i.e., what to pay attention to and how to view those things). This process of learning our native culture and its system for understanding and interacting with our world is called *enculturation* (Berry et al. 2002, 21).

### Enculturation and Credibility

From this point, the instructor must help students understand enculturation has important implications for how humans view objects in our world. First, when we encounter a new object, our native culture tells us what it is (St.Amant 2005, 79). To help students conceptualize this point, instructors can use the following example: the first time a child encounters a new item, an adult generally explains what it is and how it is used. At this point, the instructor could hold up an object (in this example, a cup), slowly point at it, and carefully repeat the following: "That is a cup. A cup. It is used for holding liquids so we can drink them." The instructor can next explain through approaches like this one that our native culture teaches us

- to look for and recognize cups among the various objects we encounter;
- to know what something must look like to be identified as a cup in our culture;
- to determine what function a cup plays in our society;
- to examine how members of our society expect a cup to be used.

The instructor should then explain, within this context, the more we are told "this object is a cup," the more we come to associate the feature of that object with what a cup should look like (Aitchison 1994, 448). Additionally, the more we see that object—that cup—used in a specific way, the more we accept the object (the cup) should or must be used that way. Thus, repeated exposure not only establishes what something is/looks like (recognizability), but it also teaches how that item should be correctly used (acceptability) (St.Amant 2005, 80–81).

Why is this important? Because it is connected to use.

### Credibility and Usability

At this point, the instructor should explain how credibility (how believable something is) and usability (if something can and should be used as directed) are connected (Connolly 1996, 29–30; Forslund 1996, 47–48). Doing so involves introducing students to a foundational concept: if something looks credible to us, we are more likely to use it (Schriver 1997, 494–95). From a technical communication perspective, this factor of use is key. After all, technical communicators want individuals to use their materials to better understand ideas (in the case of informational artifacts) or perform activities (in the case of instructional artifacts).

To help students grasp these concepts, the instructor might request students provide examples of the connections between credibility and usability. To do so, the instructor could ask students to identify an interface they all know is a credible source of information and that is recognized as easy to use (e.g., the web page for a particular university office or organization). Students would then note what features of this site's design make it credible and make it usable.

Next, the instructor would ask students how they might replicate the design of this site to create a new web page for a similar organization or office. The objective would be to create a site that will be considered equally credible and usable based upon replicating design features from the "successful" example. This approach allows the instructor to discuss how aspects of design contribute to the perceived credibility and usability of a site. It also reveals how identifying and replicating such features can be important considerations when developing parallel materials for the same audience.

*Addressing Cultural Credibility Expectations*

Once connections between credibility and usability are established, the instructor can move to aspects of culture and communication. To do so, the instructor must first note individuals from other cultures are often exposed to different artifacts over the course of their early lives. Thus, cultures can have different expectations of what a given artifact should look like (i.e., what the ideal representation of an artifact is) to be both recognizable and acceptable—or credible (St.Amant 2005, 80–81). This factor means what constitutes a credible design for a given technical communication artifact can vary from culture to culture (Keegan and Green 2003, 136–37; Kostelnick and Roberts 1998, 36–37, 348–49). These variations, moreover, can involve everything from the genres used to the content of a text to the visuals that accompany an item. So, what might be a credible technical communication artifact in one's native culture (e.g., a recognizable and acceptable instruction set) could be considered unrecognizable or inappropriate in another (St.Amant, 2006, 9–11). This disconnect affects the perceived credibility and the related use of that item in other cultural settings.

## TEACHING THE TOPIC

Teaching this topic involves reviewing materials designed for other cultures and comparing them to parallel materials designed for one's own culture. (Students, for example, might review the website Volkswagen uses to sell cars to French consumers and then compare this site to the one used to market cars to US audiences.) By reviewing sites that convey similar information to different cultural audiences, students can learn to identify patterns for how cultures share information via a common medium (e.g., organizational websites). To further explore differences in culture and design, students could compare how the French and the US sites for General Motors, Renault, and Nissan present information on a common topic (i.e., cars).

To review more general cultural communication preferences, students could compare websites on a range of topics across the same cultures. Students could, for example, compare French versions of Sony's digital camcorders sales sites, of Microsoft's Windows customer-support sites, and of Nokia's sites for purchasing mobile phones to see if any similarities in design occur. They could next look at the parallel sites each organization uses to share information on the same topic with US audiences to identify more general differences in online design between French and US websites. In so doing, students could begin to identify what design differences seem common across all sites for a given culture—or what is more general in nature—and what features seem related only to certain sites or products.

I refer to this approach as a comparative online analysis of cultures (COAC). This COAC approach allows students to engage directly with materials from other cultures and to explore concepts of international communication in almost any class in a curriculum—or across classes in a program. This is because the COAC approach does not require instructor expertise to integrate examinations of culture and communication into a given course. Rather, the challenge for instructors involves identifying the parallel online materials needed to expose students to cultural design preferences. (In this case, *parallel* means items designed to convey similar content in order to achieve similar objectives.)

*Selecting Materials*

For this COAC heuristic to be effective, instructors must identify online artifacts for comparative cultural analysis. Unfortunately, no perfect source of such materials exists. Instructors can, however, use the following process to select materials for such activities:

- Identify organizations that have developed online content for markets in other nations. Instructors should first review the websites designed by well-known multinational companies. For example, *Forbes* magazine's Global High Performers List (online at http://www.forbes.com/lists/global-high-performers-full-list.html) could serve as a resource for identifying such organizations.

- Review prospective sites to confirm they address a relatively wide range of nations and cultures. The key is using artifacts representing a range of cultures. Such diversity helps students understand the kinds of cultural communication differences that can affect international interactions. Instructors should therefore confirm that the multinationals they wish to use have parallel websites for sharing information with broad cultural groups that represent

    o Central America (e.g., Guatemala, Costa Rica, or Belize);
    o South America (e.g., Brazil, Argentina, or Suriname);
    o The Middle East or Northern Africa (e.g., Egypt, Saudi Arabia, or Turkey);
    o Southern/sub-Saharan Africa (e.g., South Africa, Senegal, or Malawi);
    o Central Asia (e.g., Pakistan, India, or Kazakhstan);
    o Eastern Asia (e.g., China, Japan, or Korea);
    o Western Europe (e.g., France, Germany, or Finland);
    o Eastern Europe (e.g., Poland, Russia, or Romania).

(Note: This listing is by no means comprehensive, but it does provide students with a diverse range of cultures for initial review.)

The idea is to assign one or more cultures/nations from each area to different students in a class. By having students share COAC findings across these cultures, instructors can help individuals understand the range of cultures and communication differences in today's global workplace.

- Determine how difficult it is to locate specific national/cultural websites within an overall site. Many multinational organizations have designed their websites to allow for easy access to a directory of global sites (i.e., websites designed to share information with specific nations). General Electric (GE), for example, uses a "Directory" page that allows individuals to easily access the websites GE has created to convey information to other nations (see http://www.ge.com/directory). Similarly, Coca-Cola has a "Country Global" page that facilitates access to a similar range of international materials (see http://www.coca-colacompany .com/). Other organizations do not. The instructor should therefore determine how difficult it might be for students to locate culture-specific materials from a common online source. (The more difficult the access to such materials, the more problematic it might be for students to engage in a COAC activity.) Instructors should then select COAC-related sources that are relatively easy to find and use.
- Identify parallel sites across the same industry. By comparing how organizations in the same industry present online information to the same cultures, students can better understand how the topic covered affects cultural design expectations. Instructors should thus try to locate two to three different multinational sites that are in the same industry and are for the same range of cultures. To examine how information in the energy sector is conveyed across cultures, for example, instructors could have students review the international websites for Exxon, Royal Dutch Shell, and British Petroleum. By comparing the websites different companies in an industry have designed for the same cultural audience, students can learn cultural communication associations for a given item or product.
- Locate parallel websites across different industries. Comparing design features within the same culture but across different industries allows students to identify more general design expectations independent of the topic discussed. (Students could, for example, review the sites Exxon, Volkswagen, and Sony have created for French audiences to see if any design approaches are common across all these sites.) Such comparisons of expectations across different industries can help students understand how the topic of a communiqué can influence cultural design expectations.

Once the instructor has identified effective materials (i.e., websites), the next step is for students to perform a COAC-based comparison and analysis of materials.

*Reviewing the Process*

Instructors can use the following process to integrate COAC approaches into technical communication classes:

- *Step 1: Reviewing the activity.* The process begins with the instructor summarizing the ideas of credibility, exposure, and culture noted in the "Literature Review" section of this chapter. The instructor would next explain that an effective way to examine cultural credibility differences involves comparing how organizations use a medium to share parallel information with different cultures. To do so, students should review the websites certain multinational organizations use to convey parallel content to customers in a range of cultures.

  The instructor can also explain that the more cultures students review and compare, the better they can understand how cultural factors affect communication expectations. To this end, the instructor should organize students into research teams and assign each team a different culture to review, analyze, and discuss via a final in-class report.

- *Step 2: Assigning cultures.* The instructor can organize students into groups of two or three (depending on the size of the class) and assign each group a particular nation/culture to review (ideally, across cultures noted in the "Selecting Materials" section of this chapter). If possible, students should review more than one culture per region to examine how communication expectations can vary across the cultures in a given area.

- *Step 3: Performing reviews.* Once student groups are assigned a culture to research, they should do a comparative review of parallel websites (i.e., sites an organization uses to share the same information with different cultures). In doing this comparison, students should identify design factors that seem common across the sites for each culture and design aspects that seem to be different when comparing the two cultures. Ideally, students could record such commonalities and differences in two side-by-side lists that would allow them to easily compare and contrast their findings.

- *Step 4: Creating checklists.* Using the lists created for step 3, students can develop a checklist for designing online materials for individuals from the culture they reviewed. This checklist would allow individuals unfamiliar with the culture to create more effective/credible online materials for sharing information with members of that culture.

  Once the checklist is complete, student groups can review the website a different multinational organization uses to share information with the same culture. The objective would be to test the checklist against other websites designed for the same audience. Students would note what aspects of their checklist appear again and what aspects do not. (The idea is to use multiple comparisons of culture-specific materials to create a more definitive, final checklist.) Students could then do additional reviews of websites other multinationals use to share online information with the same cultural group and modify their checklist with each review.

In undertaking such reviews, students should first examine multinational organizations in the same general sector (e.g., comparing the websites for Coca-Cola and Pepsi to see how similar products are marketed to the same cultures). Students can expand their review to other industries (e.g., the auto industry by using the international sites of General Motors, Mercedes, and Nissan) to examine what culture and design factors are connected to sharing information about similar topics/products (e.g., soft drinks or automobiles) and what factors seem more universal/common and are found across all sites for a culture regardless of the topic/product associated with the site. The goal is to develop more definitive design checklists that can help students better understand the communication expectations of other cultures.

- *Step 5: Sharing results.* Once this iterative review process is complete, a final class activity would ask students to compare their findings across the cultures analyzed for this assignment. To do so, each student group should give a presentation of the checklists developed from their research. Groups should note how they arrived at their final checklist and use examples from the sites they reviewed to illustrate the presence of a particular design element. Students should also compare such sites to parallel ones designed for their own culture. Such a comparison can reveal how cultural expectations associated with a topic or medium can deviate from those of the students' own culture.

This five-part COAC approach exposes students to the design expectations of a wider range of cultures. It also helps them see how cultural variations can affect communication expectations. Through such comparative activities, students can better understand how cultural communication expectations affect design choices in international contexts.

### ASSIGNMENTS

Instructors can use COAC activities to introduce concepts of culture and communication on a range of topics. This section presents two COAC-based assignments that can help students explore various aspects of culture, context, and communication. Students can do these assignments independently or in a sequence that allows them to take what they have learned in one assignment and apply it to the other.

### Comparing Expectations

The most basic COAC-related activity involves reviewing websites organizations use to share parallel information with different cultural groups. For this assignment, students could compare the website an organization has created for sharing the same information with the student's native culture and with a different culture. The objective is for students to create two design checklists: one for designing websites for their own culture and one for creating websites for the other culture.

Once the two checklists are completed, students could test them by making design predictions. To do so, students could review another set of sites used to share parallel information about a similar product with the same cultural audiences. (If, for example, students reviewed the US and the French sites for Coca-Cola and created an initial design checklist for these cultures, they would now review the US and French sites for Pepsi.) Before doing so, however, students should use the two cultural design checklists they created to predict how

- each site will be designed: they would use these checklists to make an informed guess on what aspects on these checklists they think will appear on the sites they will review;
- the two sites will differ: if they noted a major design difference between the two initial sites they surveyed (e.g., the presence of a left-hand menu bar on one but not the other), they must posit whether they expect similar variations in the new sites they will review.

Once students have generated a list of predictions for both items, they would review the actual websites for the same cultures and determine how accurate their predictions were.

Through this comparative and predictive approach, students begin to learn

- cultures can differ in the presentation expectations they have for parallel content and for sharing information through similar media (e.g., websites);
- cultures can vary in terms of the design expectations they associate with the same topic and the same medium;
- expectations related to conveying information on a common topic online can vary within a culture.

Instructors can then expand on this initial assignment by asking students to do one or more of the following activities:

- Compare how other (e.g., three, four, or five) organizations in the same area (e.g., soft drinks, in the case of the Coke-Pepsi example used here) share parallel information with the same cultural groups. Such activities can help students better understand what communication/design aspects the members of a culture associate with sharing information on a topic via online media.

- Compare how organizations in different sectors (i.e., areas or industries other than those examined in the initial comparison) share parallel online information on different topics with the same culture (i.e., organizations selling different products—say, automobiles versus the soft-drink companies examined originally). Such activities can help students better see what aspects of culture and design are connected to communicating in a given medium (e.g., a website) versus communicating about a specific topic in a common medium.
- Compare how the same companies design sites for sharing parallel information with other cultures (i.e., cultures beyond those the students have examined up to this point). Such activities can help students see how cultures can vary in expectations for sharing information on a common topic via the same medium.

Through such comparisons, students gain an initial understanding of differences in cultural communication expectations.

### Localizing Materials

Localization involves taking materials designed for one cultural group and revising them to meet the communication expectations of another. A COAC approach can help familiarize students with the dynamics of such processes.

In this case, students would do an initial COAC review to create foundational design checklists for members of their own culture and for individuals from another culture. This time, however, students could use a different approach when reviewing a second set of websites other organizations use to share parallel information on the same topic. When reviewing the website for a different organization, students would only review the site designed for members of their own culture. They would use the design checklist they created to make suggestions on how the new/second site designed to share information with members of their same culture should be revised to convey information to members of the other culture they studied earlier. (If France was the other culture students reviewed, they would use the design checklists they created to make suggestions on how to revise the new US site for French users.)

Once students record their localization suggestions, they can review the actual site the organization uses to share information with that other culture and compare their predictions to what they actually find. Students can use their review of this second site to modify their design checklist accordingly. From this point, instructors can ask students to engage in activities to further explore ideas of culture and communication, including

- asking students to use this second example to refine their design checklists and then review how a third organization uses online media to share information about that same topic with members of their own culture. Again, students would use their (now revised) checklist to predict how to localize that new site for members of the other culture they are studying and then compare their design suggestions to the actual, localized version of the site. Students could repeat this process over several iterations and continue to modify their design checklist. Through such a process, students can better understand how localization works and how cultures can associate a common set of design expectations with communicating information about a particular topic.

- requiring students to use the checklists they designed to localize different web content for the same cultural audience. For example, the initial COAC review and follow-up localization activity could focus on how organizations share parallel information on the same topic with cultural audiences (e.g., how Coca-Cola and then Pepsi share information on soft drinks with users in the United States and in France). Students could then be asked to review the US website Microsoft uses to share information with US audiences and—using the checklist they have designed via reviews of the Coca-Cola and Pepsi sites—suggest how to localize the US-targeted Microsoft site to share information with French audiences. Next, students would review the actual Microsoft France site to see how accurate their localization predictions are. This approach could be repeated across a range of industries for the same cultures. Through such comparisons, students begin to gain an understanding of what localization aspects seem connected to cultural expectations of a medium in general.

These activities help students further explore how cultural expectations can influence the design of materials.

### The Next Steps

The two kinds of assignments described here create a foundation from which students can explore cultural aspects of design from a range of perspectives. As such, instructors should feel free to modify these assignments to address the objectives of a given class. (In a class on policy writing, for example, instructors could use the COAC approach to review how different national governments design websites to share information and updates on various policies with citizens.)

Instructors could also ask students to take a more active role in selecting materials for use in a COAC review. The instructor, for example, could assign students to locate the websites different multinational organizations use to share information with different cultural audiences. The instructor would then have students share these examples with classmates and use them for different in-class COAC-related activities. Such sharing might include requiring students to find other multinational organizations the class can use to expand its COAC activities and explore how organizations in the same industry share information with different cultures.

In the end, the goal of such COAC-related assignments and activities is to foster engagement. Thus, the more instructors can do to get students engaged with and interacting with online materials, the more such experiences can enhance student understanding of cultural differences in communication expectations.

## CONCLUSION

Cultural factors can influence communication expectations in nuanced and unexpected ways. As workplaces become increasingly international in nature, instructors must emphasize the importance of understanding how to work in cross-cultural contexts. Doing so involves helping students develop a foundational understanding of how cultural factors can affect communication expectations. The framework, approach, and activities described in this chapter represent one way technical communication instructors can accomplish this goal.

By requiring students to examine materials used to share information with various nations, instructors can raise student awareness of cultural communication differences. By asking them to compare such materials to those of their own culture, instructors can increase student understanding of the nuanced nature of cultural communication expectations. Such approaches also help students develop mechanisms they can use to continue to explore culture and communication factors beyond the classroom.

In truth, the teaching of culture and communication is a complex undertaking. Effective education, however, is a matter of inspiring students to engage actively with and to want to learn more about a topic. The COAC approach provides students with a mechanism that can engage and inspire them to further their studies of culture and communication both during their formal studies and after graduation. By integrating this approach into the different courses across a curriculum, instructors can provide students with a range of opportunities to test what they have learned and establish and expand their understanding of and appreciation for cultural aspects of technical communication.

## DISCUSSION QUESTIONS

1. How do you see the ideas examined in this chapter as guiding the choices you might make about when and how to integrate aspects of culture and communication into the classes you teach? How might these factors affect the way(s) in which you approach the process of developing new classes in the future?

2. The process of reviewing international materials for use in the kinds of activities described here can be quite time consuming. What suggestions do you have—or what approaches would you suggest others use—to keep this review process manageable but effective in terms of selecting materials for such activities?

3. Should the instructors who wish to use this approach to teaching culture and communication also engage in additional research to learn more about cultural communication expectations in general? Why or why not? Do you think it is problematic for instructors to attempt to teach cultural aspects of communication without first engaging in more research on the topic themselves? Why or why not?

4. What do you think the greatest challenges might be to using the approach and activities noted here to integrate the teaching of culture and communication into different technical communication classes? Do you think these challenges are universal to integrating the teaching of this topic in all the classes in a program, or would it be easier to do with certain types of classes? Why or why not?

5. The approach presented here teaches students about culture and communication factors without asking students to actually interact directly with or observe individuals from other cultures. Do you think this lack of interaction with other cultures is problematic in relation to teaching this topic in technical communication classes? Why or why not?

*References*

Aitchison, Jean. 1994. "Bad Birds and Better Birds: Prototype Theory." In *Language: Introductory Readings*, 4th ed., edited by Virginia P. Clark, Paul A. Eschholz, and Alfred F. Rosa, 445–59. New York: St. Martin's.

Berry, John W., Ype H. Poortinga, Marshall H. Segall, and Pierre R. Dasen. 2002. *Cross-Cultural Psychology: Research and Applications*. 2nd ed. New York: Cambridge University Press.

Connolly, John. 1996. "Problems in Designing the User Interface for Systems Supporting International Human-Human Communication." In *International User Interfaces*, edited by Elisa M. del Galdo and Jakob Nielsen, 20–40. New York: John Wiley & Sons.

Forslund, Charlene Johnson. 1996. "Analyzing Pictorial Messages Across Cultures." In *International Dimensions of Technical Communication*, edited by Deborah C. Andrews, 45–58. Arlington, VA: Society for Technical Communication.

Holliday, Adrian. 2013. *Understanding Intercultural Communication: Negotiating A Grammar of Culture*. New York: Routledge.

Keegan, Warren J., and Mark C. Green. 2003. *Global Marketing*. 3rd ed. Upper Saddle River, NJ: Prentice Hall.

Kostelnick, Charles, and David D. Roberts. 1998. *Designing Visual Language: Strategies for Professional Communicators*. Boston, MA: Allyn and Bacon.

Ross, Brian H., and Valerie S. Makin. 1999. "Prototype versus Exemplar Models in Cognition." In *The Nature of Cognition*, edited by Robert J. Sternberg, 205–41. Cambridge: MIT Press.

Schriver, Karen A. 1997. *Dynamics in Document Design: Creating Text for Readers*. New York: John Wiley & Sons.

St.Amant, Kirk. 2005. "A Prototype Theory Approach to International Website Analysis and Design." *Technical Communication Quarterly* 14 (1): 3–91.

St.Amant, Kirk. 2006. "Globalizing Rhetoric: Using Rhetorical Concepts to Identify and Analyze Cultural Expectations Related to Genres." *Hermes—Journal of Language and Communication Studies* 37: 47–66.

# ABOUT THE AUTHORS

TRACY BRIDGEFORD is professor of technical communication at the University of Nebraska at Omaha. She cofounded and coedited *Programmatic Perspectives*. She serves as a reviewer for *Kairos* and is currently on the editorial board of *Technical Communication Quarterly*. In 2015, she published, with Kirk St.Amant, *Academy-Industry Relationships: Perspectives for Technical Communicators*. In 2014, she published, with Karla Saari Kitalong and Bill Williamson, *Sharing Our Intellectual Traces: Narrative Reflections from Administrators of Professional, Technical, and Scientific Programs*. She has contributed chapters to *Resources in Technical Communication: Outcomes and Approaches* and *Teaching Writing with Computers: An Introduction*, and *Innovative Approaches to Teaching Technical Communication*, which she also coedited. She has also published in *Kairos: A Journal of Rhetoric, Technology, and Pedagogy*. She coedited a special issue of *Technical Communication Quarterly* on *Techne* and technical communication.

PAM ESTES BREWER is a technical communicator, educator, and management consultant. She teaches in Mercer University's School of Engineering and directs the MS in technical communication management program. She researches and trains on remote teaming, and her book, entitled *International Virtual Teams: Engineering Global Success*, was published in 2015. Join her on LinkedIn in the Remote Teaming Network group and follow her on Twitter @brewerpe.

EVA BRUMBERGER is an associate professor and head of the technical communication program at Arizona State University. Her research interests include visual rhetoric and document design, intercultural communication, workplace practices, and pedagogy. She has published in a variety of journals, has coedited a collection on teaching visual communication, and serves on the editorial boards of the *Journal of Visual Literacy*, *Communication Design Quarterly*, and *Business and Professional Communication Quarterly*.

DAVE CLARK is associate dean for the humanities and an associate professor in the Department of English at the University of Wisconsin–Milwaukee. His research focuses on content management, information design, and the rhetoric of technology. Dave studies and teaches the rhetoric of technology, with special research interests in content management and controlled language. His work appears in *IEEE Transactions on Professional Communication*, *Technical Communication Quarterly*, *Communication Design Quarterly*, and the *Journal of Business and Technical Communication*.

PAUL DOMBROWSKI is a professor of English at the University of Central Florida, where he has taught the rhetorical and ethical aspects of technical and scientific communication and the history and theory of rhetoric for over fifteen years. His PhD is in communication and rhetoric from Rensselaer Polytechnic Institute. His special interests are rhetoric and ethics in relation to space exploration, military technologies, and nuclear weapons. He has chaired the Ethics Committee of the Association of Teachers of Technical Writing since its formation over fourteen years ago. His principal published works are two books: *Ethics in Technical Communication* (Allyn and Bacon, 2000) and *Humanistic Aspects of Technical Communication* (Baywood, 1994).

JAMES M. DUBINSKY is associate professor of rhetoric and writing in the Department of English at Virginia Tech (VT). From 1998 until 2007, Jim was the founding director of the professional writing program, and from 2008 to 2011, after the shooting tragedy, he served as founding director of what is now VT-ENGAGE. Much of his scholarship focuses on civic engagement and the scholarship of teaching and learning. Jim is also a veteran, having served in the US Army on active duty from 1977 to1992 and in the reserves from 1992 to 2004 before retiring as a lieutenant colonel. His current projects include creating a Veterans in Society program at VT and working on a community-based dialogue on race. Jim also serves VT as the chair of the newly founded Veterans Caucus. He has served the Association for Business Communication (ABC) as an elected officer (2006–2010) and since July 2011 as the executive director.

PETER S. ENGLAND is an instructional associate professor in the Zachry Department of Civil Engineering at Texas A&M University. While he continues to be interested in technical and engineering communication research, his primary duty is teaching engineering communication to senior civil engineering majors.

DAVID K. FARKAS is a professor emeritus in the Department of Human Centered Design & Engineering at the University of Washington. He remains active in the fields of technical communication and information design. Dave has published in the areas of software-user assistance (computer documentation), slideware/PowerPoint, web design, writing and text design, manuscript editing, and other topics. A special interest is QuikScan, a document format that employs within-document summaries to improve retention and allow readers to choose their preferred level of detail. His publications are available at ResearchGate, www.quikscan.org, and http://faculty.washington.edu/farkas. Dave received a BA in English from the University of Rochester and a PhD in language and literature from the University of Minnesota. He began his teaching career at Texas Tech University and West Virginia University.

BRENT HENZE is associate professor of technical and professional communication at East Carolina University, where he serves as professional communication graduate advisor and internship program coordinator and teaches courses in technical editing, grant writing, science writing, communication research, and genre. He has served as associate editor for *Technical Communication Quarterly*. His publications include a coauthored monograph on the history of writing instruction and other publications on technical communication, field-based learning, scientific writing, and genre theory and research.

THARON W. HOWARD teaches in the master of arts in professional communication program and the rhetorics, communication, and information design doctoral program at Clemson University. He is a recipient of the STC's Jay R. Gould Award for Excellence in Teaching Technical Communication and is a nationally recognized leader in the field of usability and user-experience research. As director of the Clemson University Usability Testing Facility, he has conducted sponsored research aimed at improving and creating new software interfaces, online document designs, and information architectures for clients including Pearson Higher Education, IBM, NCR Corp., and AT&T. For his work promoting the importance of usability in both industry and technical communication, Dr. Howard was awarded the Usability Professionals Association's Extraordinary Service Award. Howard is the series editor for the Routledge-ATTW Series in Technical and Professional Communication and he also serves as the production editor for Clemson's Center for Electronic and Digital Publishing where—in addition to producing scholarly journals, books, fliers, and brochures—he teaches MAPC and RCID graduate students to create and maintain digital publications and websites. He also designed and directed Clemson's Multimedia Authoring Teaching and Research Facility, where faculty and graduate students in architecture, arts, and humanities learn to develop fully interactive, stand-alone multimodal productions and experiment with emerging instructional technologies, augmented-reality devices, and interface designs. Howard is the author of *Design to Thrive: Creating Online Communities and Social Networks That Last* and *A Rhetoric of Electronic Communities;* coauthor of *Visual Communication: A Writer's Guide;* and coeditor of *Electronic Networks: Crossing Boundaries and Creating Communities.* He has also authored articles in journals including *Technical Communication, Technical Communication Quarterly,* and *Computers and Composition.*

DAN JONES is a professor of English at the University of Central Florida in Orlando. He has taught a variety of undergraduate and graduate technical communication courses since 1980, and he has provided over thirty technical communication workshops on the essentials of technical communication to professionals from numerous companies. His books include *Technical Writing Style* (Allyn and Bacon, 1998), *The Technical Communicator's Handbook* (Allyn and Bacon, 2000), and *Technical Communication: Strategies for College and the Workplace* (Longman, 2002), coauthored with Karen Lane. He received the Jay R. Gould Award for Excellence in Teaching Technical Communication from the Society for Technical Communication (STC) in 1998, was elected an STC Fellow in 2000, and received the Ronald S. Blicq Award from IEEE's Professional Communication Society in 2003.

KARLA SAARI KITALONG is professor of humanities at Michigan Technological University and director of the program in scientific and technical communication. Her research and teaching interests are situated at the intersections of visual rhetoric and usability. Her work in formative evaluation and user-centered design of mixed-reality simulations, humanities learning games, and immersive databases has been funded by both the National Science Foundation and the National Endowment for the Humanities.

TRACI NATHANS-KELLY, currently teaches in Cornell University's College of Engineering. As a member of the engineering communications program, she interacts daily to help undergraduates and graduates alike to hone their messages to their instructors, internship managers, co-op directors, thesis and dissertation committees, and audiences at conferences. Before coming to Cornell in 2012, she spent fourteen years working at the University of Wisconsin–Madison in the College of Engineering. Mostly notably, she taught in the online masters of engineering management (MEM) and the masters of engine systems (MEES) degree programs, aiding practicing engineers in communicating targeted technical messages with strength, evidence, and conviction. In 2017, she finished her tenure as the series editor for the IEEE professional communication book series titled Professional Engineering Communication. The book, *Slide Rules: Design, Build, and Archive Presentations in the Engineering and Technical Fields,* was coauthored with Christine G. Nicometo. Aside from campus teaching, she conducts workshops and training for such entities as The Boeing Company, Flad Architects, IEEE-USA, and Wolters Kluwer.

CHRISTINE G. NICOMETO, hhas shared her passion for presentations with students and professionals worldwide through her teaching, consulting, and speaking career. As a faculty member in the University of Wisconsin–Madison's master of engineering online programs, she has taught practicing engineers around the globe for well over a decade. Partnering with Dr. Traci Nathans-Kelly, she coauthored their book, *Slide Rules: Design, Build, and Archive Technical Presentations* (2014). Their work on presentation design and online presentation pedagogy has been influential to the careers of countless professionals in business, industry, and academia. Christine currently directs the foundations of professional development online graduate certificate program for the University of Wisconsin–Madison and regularly consults in industry on technical communication professional development. She is an active member of IEEE and ASEE and coaches the next generation of technical presenters in her daughter's FIRST Lego League robotics team. When she is not working, she enjoys sailing and traveling with her family as often as she can.

KIRK ST.AMANT is a professor and the Eunice C. Williamson Endowed Chair in Technical Communication at Louisiana Tech University, and he is also an adjunct professor of international health and medical communication with the University of Limerick (Ireland). His main research interests are international communication and information design for global audiences, with a particular focus on the globalization of online education and health and medical communication for international audiences. He has taught online and hybrid courses for universities in Belize, Denmark, Finland, Ireland, Ukraine, and the United States.

# INDEX

**G**

gender bias, 55

genre, 13–14, 54; of procedures, 122, 125, 136–37, 226

genre: and analysis, 15, 83, 85–86; choice in, 78–79, 84–85; and constraints, 79; and context of use, 74–77, 82; definitions of, 73; ecologies, 80; expectations, 30; flexibility of, 78; and form, 81–82; interdependence of, 80; recurrence of, 74–75; regularity in, 76, 78; repertoires, 80; rules, 19, 72, 78–79, 81–82; and sequences, 80; and sets, 80; as social, 73, 75, 76, 77, 86; as social action, 73, 75. *See also* rhetorical situation

genre systems, 80

genre theory, 73; archaeological model, 77; ecological model, 77, 82

genres of practice, 22

Gestalt, 108

goals, 111, 114, 116, 119, 120

Gore, Al, 145

grading, 117

graphical representations, 90

graphical user interface (GUI). *See* user interface

graphics, for slides, 214–215

graphs, 98

grouping, 108–111, 114, 115

guest instructors, 136

**H**

Habermas, Jurgen, 142

habits, 107

*Harvard Business Review,* 58

Hawthorne Effect, 198

Headers/headings, for slides, 212–217

headings, 107, 111, 116, 120

help systems. *See* online help systems

hierarchy, 111; of authority or power, 54

"how-to" discourse. *See* procedures

Hudson, Randolph, 43

human factors for technical communicators (COE), 126

humor, 54

**I**

idea generation, 107

*IEEE Transactions on Professional Communication,* 60

illustrations, 95

images, 110, 113, 116

impatience, 126

imperative voice. *See* steps

impersonal style, 52

importance of style choices, 45

imprecise or ambiguous language, deliberate, 55

industrial engineers, 49

industry practices in our classrooms, 180

industry "malpractices," 180

information access, 129

information communication technologies (ICFS), affordances of, 166–67

information design, 7–9, 105–106

information graphics, 14, 90; static, 101; interactive, 101

information needs. *See* user needs

information usefulness, 184

informed consent, 187, 199–200

institutional review board, 187

instructional manual, 181

instructions, 114, 115, 187–190, 195; as synonym for procedures, 122. *See also* procedures

integrate usability testing in the document-design process and classrooms, 176–177

integrating design, 106, 112, 116, 119

intelligent content, 63

interactivity, 126

international and intercultural communication, 16

interpretation of graphics, 94, 100

introductory paragraph of procedures. *See* paragraphs

introductory technical communication courses, 45

introductory technical communication textbooks, 44, 45

intuition, 106

invention, 73

invention heuristics, 180

ISO 9241-11, definition of usability, 182

issues of credibility, 11

issues of style, 42

italics, 25

**J**

jargon, 51; undefined, 50

job advertisements (job ads), 28, 32

job application, 29, 113

**K**

Katrina, 2

Katz, Stephen, 10

keeping subjects and verbs close together, 45